Lecture Notes in Applied and Computational Mechanics

Volume 31

Series Editors

Prof. Dr.-Ing. Friedrich Pfeiffer
Prof. Dr.-Ing. Peter Wriggers

Lecture Notes in Applied and Computational Mechanics

Edited by F. Pfeiffer and P. Wriggers

Further volumes of this series found on our homepage: springer.com

Wave Propagation in Infinite Domains

With Applications to Structure Interaction

Lutz Lehmann

Dr.-Ing. Lutz Lehmann
Institut für Angewandte Mechanik
Technische Universität Braunschweig
Spielmannstraße 11
38106 Braunschweig
Germany
lutz-o.lehmann@tu-bs.de

Library of Congress Control Number: 2007923193

ISSN   1613-7736

ISBN   978-3-540-71108-7   Springer Berlin Heidelberg New York

Springer is a part of Springer Science+Business Media
springer.com
© Springer-Verlag Berlin Heidelberg 2007

Typesetting: Integra Software Services Pvt. Ltd., India

Cover design: WMXDesign GmbH, Heidelberg

Printed on acid-free paper   SPIN: 11959489   42/3100/Integra   5 4 3 2 1 0

The present book has been accepted as Habilitation Thesis from the Department of Civil Engineering of the Technical University Carolo Wilhelmina at Braunschweig. The venia legendi for the scientific area of "Mechanics" was given on 14th of July 2006.

Contents

Part II Applications

Introduction

The goals of this work are to present the theoretical fundamentals and to demonstrate the applications of a new numerical model which has the ability to simulate the *wave propagation in infinite domains*.

Wave propagation

Wave motions are fascinating physical phenomena with important practical applications. Physicists and engineers are interested in the reliable simulation of processes in which waves propagate in solids or fluids. The effect of waves is familiar to everyone, e.g., transmission of sound or radio waves, vibrations due to traffic or construction work, or water waves. From all wave types in nature, here, attention is focused only on linear waves in ideal fluids (e.g., acoustics) and elastic domains. Acoustic waves (sound) are small oscillations of pressure in a compressible ideal fluid (acoustic medium). These oscillations interact in such a way that energy is propagated through the medium. The governing equations are obtained from fundamental laws for compressible fluids, whereas in an elastic medium, waves propagate in the form of oscillations of the stress field. This work deals with theoretical and practical issues arising in the computational simulation of wave propagation based on scalar and vector wave equations, as well as fluid-structure interaction and soil-structure interaction. While each class of problems has its distinctive features, there are underlying similarities in the mathematical models, leading to similar numerical effects in computational implementations.

Infinite domain

Unbounded domains are often encountered in scientific and engineering applications. Examples are radar and sonar technology, wireless communication, aero acoustic, fluid dynamics, earthquake simulations, or even quantum chemistry. Typically the phenomenon of interest is local but embedded in a vast surrounding medium. Although the exterior region is not truly unbounded, the boundary effects are often negligible, so that one further assumption simplifies the problem by replacing the vast exterior by an infinite medium. The dynamical behaviour of the infinite space itself could be reproduced (with more or less accuracy) by so-called absorbing boundary conditions.

Numerical model

The introduced numerical model to handle such classes of problems is based on a coupled *finite element/scaled boundary finite element* (FE/SBFE) algorithm. For the near-field mapping the commonly used finite element method (FEM) is applied. The benefit of the FEM is founded among other things on its adaptability to given boundary conditions and the possibility to include physical and geometrical non-linearities. But finite element methods have been conceptually developed for the numerical discretisation of problems on *bounded* domains. Their application to unbounded domains involves domain decomposition by introducing an artificial boundary around the structure or domain of interest. At the artificial boundary, the finite element discretisation can be coupled in various ways to some discrete description of the analytical solution. Some of those coupling approaches are based on the series representation of the exterior solution, e.g., localised Dirichlet-to-Neumann and other absorbing boundary conditions, perfectly matched layer method or infinite elements. Unacceptable wave reflections at the boundary of the finite element discretised domain are removed by a coupling with the scaled boundary finite element method (SBFEM) which was developed by Wolf and Song [148]. Under the immense variety of non-reflecting boundary conditions the scaled boundary finite element method was chosen, because it has some unique features:

- reduction of the spatial dimension by one without requiring a fundamental solution,
- no discretisation of free and fixed boundaries and interfaces between different materials,
- influence of the infinite far-field could be stored in the form of influence matrices for further simulations (different near-field geometries or load cases), and
- straightforward incorporating in existing finite element code and concise coupling with finite elements.

Initially, to achieve a scaled boundary formulation one has to discretise the boundary with finite elements. Applying the scaled boundary transformation and Galerkin's weighted residual method, the governing partial differential equations are reformulated as the SBFE equation in displacement (solids) or pressure (fluids) with an introduced radial coordinate as independent variable. The coefficient matrices of these SBFE equations are calculated and assembled similarly as the stiffness (or compressibility) and mass matrices of finite elements. Introducing the definition of the dynamic stiffness, a system of non-linear first-order ordinary differential equations with the frequency as independent variable is obtained. For a time domain analysis a further transformation is mandatory. A final numerical discretisation allows a computer based analysis.

Outline of content

This work is split into part I - the discussion of the theoretical fundamentals - followed by the exemplary presentation of three fields of application in part II.

The first part will give a short overview of the physical governing equations and the theoretical background of FEM and BEM. The focal point here is the formulation of the elastodynamic equations in the time domain, for a detailed description an extensive number of monographs is released. In contrast, the theoretical fundamentals of the SBFEM will receive a detailed presentation in which the derivation of virtual work will be introduced.

In order to accomplish practical simulations in the time domain, the SBFEM has to be improved with respect to efficiency. The non-locality in space and time inhibits the handling of realistic problem cases. Different possibilities of performance improvement are pictured which notably reduce - especially in combination with each other - the numerical effort and which make the coupled model useful for practical application.

Initially a recursive algorithm is introduced which allows a quasi-linear dependency of calculation time on the number of calculated time-steps. Furthermore, the unit-impulse response matrices which are essential for the analysis in the time domain can be generated sparsely populated in opposition to fully populated matrices. This leads to additional reduction of CPU time consumption and further to a decrease in memory requirements. In closing, the dependency of computational effort on the number of degrees of freedom at the near-field/far-field interface can be reduced from quadratic to quasi-linear ratio by application of hierarchical matrices. The combination of all mentioned proceedings leads to an efficient computer code.

Benchmark examples show that the introduced approximations do not lead to any loss of accuracy. Furthermore, comparisons with the Boundary element method (BEM) or the coupled FE/BE model respectively show that there are no major deviations to the results of the FE/SBFE method. From a multiplicity of possible applications, three fields are chosen for this work:

- wave propagation in fluids (acoustics),
- the simulation of dynamic behaviour of offshore wind turbines,
- and finally the seismic analysis of a multi-storey building with regard to soil-structure interaction.

For all examples evaluations in the time domain are executed which are complemented by calculations in the frequency domain for individual problem cases.

The simulation of wave propagation in fluids includes an investigation of a sound barrier and a time domain analysis of a waveguide. For analysing an offshore wind turbine the extensive description of the offshore environment is essential. Here, time-dependent loads arising from different sources, such as wind, waves, rotor-dynamics or ice loadings, dominate the dynamical behaviour. Furthermore, the soil-structure interaction – where simple constitutive models for saturated sand and clay are used – is of notably interest. Finally, an earthquake engineering example is added where again the soil-structure interaction has to be taken into account.

Part I

Theory

1. Finite element method

The finite element method (FEM) can be viewed as an application of the Rayleigh-Ritz method and is mostly based on a displacement approximation, see, e.g. [11]. It is based upon subdividing the deformable body or the structure in several finite elements, see Fig. 1.1, interconnected at nodes on the element edges. These elements have mostly simple geometries and a special structural function, such as plate or shell elements.

In the following, a short overview of the FEM will be given (detailed descriptions are widely available, see, e.g. Bathe [13] or Zienkiewicz [153]).

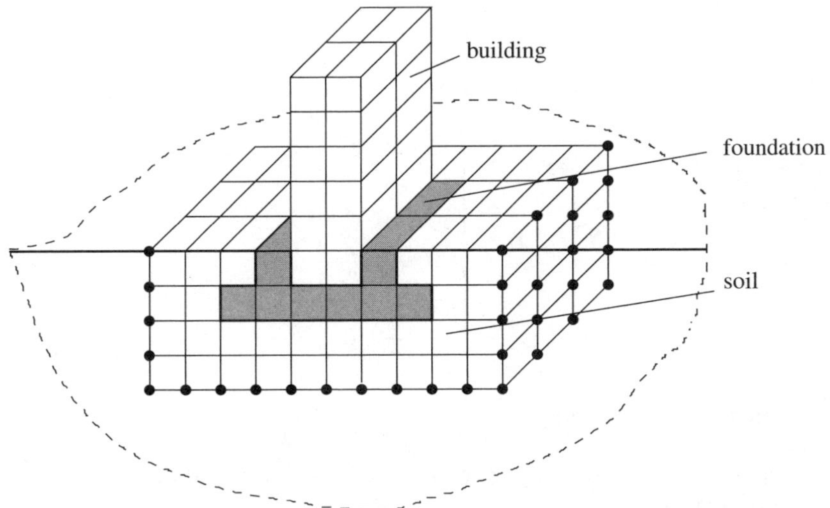

Fig. 1.1. Three-dimensional discretisation of a soil-structure interaction problem

A major strength of the FEM is the ease with which complex geometries, anisotropic materials, and boundary conditions can be handled. Also advantageous is the possibility to include non-linearities in the numerical simulation.

1.1 Governing equations

A summary of the governing equations of linear-elastic theory , including the equations of motion, is given for three-dimensional elastodynamics. Here, the constitutive equations of linear theory of elastic materials are repeated briefly. Detailed explanations can be found, e.g., in the books of Gould [61], or Eringen and Suhubi [49].

1.1.1 Strain-displacement relationship

For linear theory, the strain tensor is defined by the strain-displacement relationship:

$$\boldsymbol{\varepsilon} = \mathbf{D}\mathbf{u} \, , \tag{1.1}$$

with the strain tensor $\boldsymbol{\varepsilon}$ in pseudo-vector notation

$$\boldsymbol{\varepsilon} = \begin{bmatrix} \varepsilon_{11} \\ \varepsilon_{22} \\ \varepsilon_{33} \\ 2\varepsilon_{23} \\ 2\varepsilon_{31} \\ 2\varepsilon_{12} \end{bmatrix} , \tag{1.2}$$

and the gradient operator $\mathbf{D}$

$$\mathbf{D} = \begin{bmatrix} \frac{\partial}{\partial x_1} & 0 & 0 \\ 0 & \frac{\partial}{\partial x_2} & 0 \\ 0 & 0 & \frac{\partial}{\partial x_3} \\ 0 & \frac{\partial}{\partial x_3} & \frac{\partial}{\partial x_2} \\ \frac{\partial}{\partial x_3} & 0 & \frac{\partial}{\partial x_1} \\ \frac{\partial}{\partial x_2} & \frac{\partial}{\partial x_1} & 0 \end{bmatrix} . \tag{1.3}$$

1.1.2 Constitutive equation

The components of the stress tensor $\boldsymbol{\sigma}$ are combined via Hooke's law with the vector of strain states $\boldsymbol{\varepsilon}$ as

$$\boldsymbol{\sigma} = \mathbf{E}\boldsymbol{\varepsilon} \, , \tag{1.4}$$

with $\boldsymbol{\sigma}$:

$$\boldsymbol{\sigma} = \begin{bmatrix} \sigma_{11} \\ \sigma_{22} \\ \sigma_{33} \\ \sigma_{23} \\ \sigma_{31} \\ \sigma_{12} \end{bmatrix} , \tag{1.5}$$

and the elastic modulus matrix $\mathbf{E}$:

$$\mathbf{E} = \begin{bmatrix} E_{11} & E_{12} & E_{13} & E_{14} & E_{15} & E_{16} \\ & E_{22} & E_{23} & E_{24} & E_{25} & E_{26} \\ & & E_{33} & E_{34} & E_{35} & E_{36} \\ & & & E_{44} & E_{45} & E_{46} \\ & & & & E_{55} & E_{56} \\ \text{sym.} & & & & & E_{66} \end{bmatrix}, \tag{1.6}$$

where the elastic modulus matrix may be stated for anisotropic material.

1.1.3 Dynamic equilibrium

The dynamic equilibrium in the time domain, also called equation of motion, is given by:

$$\mathbf{D}^T\boldsymbol{\sigma}(t) + \mathbf{f}(t) - \kappa\dot{\mathbf{u}}(t) - \rho\ddot{\mathbf{u}}(t) = \mathbf{0}. \tag{1.7}$$

Now, all states, i.e., the stress $\boldsymbol{\sigma}(t)$, the body load $\mathbf{f}(t)$, the displacement $\mathbf{u}(t)$, velocity $\dot{\mathbf{u}}(t)$, and acceleration vector $\ddot{\mathbf{u}}(t)$ are time-dependent. ρ is the density and κ is the damping property parameter .

1.1.4 Boundary conditions

At all points of the boundary, applicable boundary conditions must be specified, either displacements or surface tractions:

$$\mathbf{u} = \bar{\mathbf{u}} \text{ on } \Gamma_u, \tag{1.8}$$

where $\bar{\mathbf{u}}$ are the prescribed displacements, and

$$\mathbf{t} = \bar{\mathbf{t}} \text{ on } \Gamma_t, \tag{1.9}$$

where $\bar{\mathbf{t}}$ are the prescribed tractions.

1.2 Virtual work formulation

The task is to find a displacement field $\mathbf{u}$ which satisfies the equilibrium equation (1.7) in the domain and the conditions (1.8) and (1.9) on the boundary.

An alternative formulation of the equilibrium requirement is the virtual work statement. Using $\delta\mathbf{u}$ to represent a virtual displacement, and

$$\delta\boldsymbol{\varepsilon} = \mathbf{D}\delta\mathbf{u} \tag{1.10}$$

to represent the corresponding virtual strains, the virtual work equation reads:

$$\underbrace{\int_V \delta\boldsymbol{\varepsilon}^T \boldsymbol{\sigma}\, dV}_{\text{internal work}} \;-\; \underbrace{\int_V \delta\mathbf{u}^T (\kappa\dot{\mathbf{u}} + \rho\ddot{\mathbf{u}})\, dV}_{\text{work of internal loads}}$$

$$-\; \underbrace{\int_V \delta\mathbf{u}^T \mathbf{f}\, dV}_{\text{external work of body loads}} \;-\; \underbrace{\int_\Gamma \delta\mathbf{u}^T \mathbf{t}\, d\Gamma}_{\text{external work of body tractions}} \;=\; 0 \,. \tag{1.11}$$

If this equation is satisfied only for a subset of virtual displacement fields, equilibrium is satisfied in a weak sense.

1.3 Approximative solution

The FEM seeks an approximate solution $\tilde{\mathbf{u}}$ for $\mathbf{u}$, using interpolation techniques. The domain under consideration Ω is divided into finite elements and a set of m interpolation nodes $\mathbf{z}^1, \ldots, \mathbf{z}^m$ (in global numbering) is chosen. A given finite element E_e contains n such nodes $\mathbf{z}^i$ ($1 \leq i \leq n$, in local numbering), associated with n interpolation functions $N^i = N^i(\eta_1, \eta_2, \eta_3) = N^i(\boldsymbol{\eta})$, where $\boldsymbol{\eta}$ denotes the local coordinate system of the finite element (see Fig. 1.2). Then, the approximation $\tilde{\mathbf{u}}$ for a single element can be written as:

$$\tilde{\mathbf{u}}(\mathbf{x},t) = \sum_{i=1}^{n} N^i(\boldsymbol{\eta})\mathbf{u}^i(t) = \mathbf{N}\tilde{\mathbf{u}}(t) \,, \tag{1.12}$$

where $\mathbf{x}$ and $\boldsymbol{\eta}$ are related by the mapping

$$\mathbf{x}(\boldsymbol{\eta}) = \sum_{i=1}^{n} N^i(\boldsymbol{\eta})\mathbf{x}^i, \quad \boldsymbol{\eta} \in \Delta_e \tag{1.13}$$

and $\mathbf{u}^i(t) = \tilde{\mathbf{u}}(\mathbf{z}^i,t)$ are the nodal values of the approximation $\tilde{\mathbf{u}}$ of $\mathbf{u}$.

In the standard finite element approach, the shape functions have unit value at a particular node and zero value at all other nodes. In this case, $\tilde{\mathbf{u}}(t)$ can be identified as the time-dependent nodal displacements . Shape functions are defined in a local or *intrinsic* coordinate system (η_1, η_2, η_3) which is different from the global Cartesian coordinate system (x_1, x_2, x_3) in which, e.g., the external forces and displacements of the assembled system will be measured. Thus, a different coordinate system is used for every element. Clearly, it is necessary to transform the coordinates of the displacements, velocity, acceleration and force components before an assembly of all elements to the complete system is performed.

The strains associated with the approximate displacement field are:

$$\tilde{\boldsymbol{\varepsilon}}(\mathbf{x},t) = \mathbf{D}\tilde{\mathbf{u}}(\mathbf{x},t) = \mathbf{DN}\tilde{\mathbf{u}}(t) \,. \tag{1.14}$$

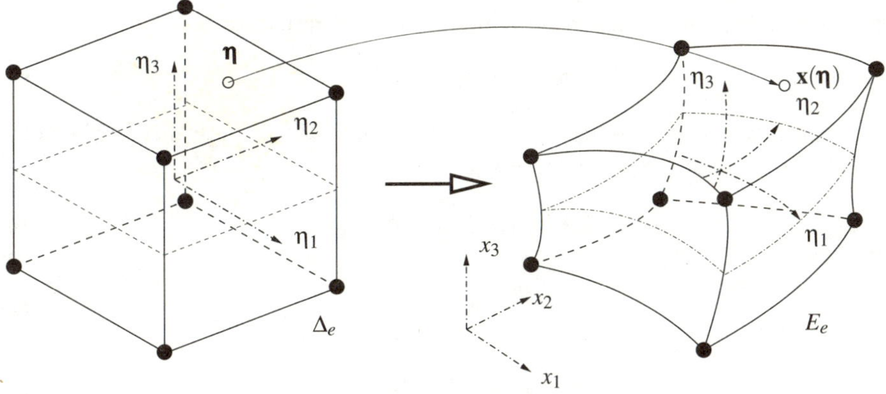

Fig. 1.2. Mapping of a three-dimensional parent element Δ_e to an eight-node finite element E_e

The approximate stresses follow (1.4):

$$\tilde{\boldsymbol{\sigma}}(\mathbf{x},t) = \mathbf{E}\tilde{\boldsymbol{\varepsilon}}(\mathbf{x},t) = \mathbf{E}(\mathbf{DN})\tilde{\mathbf{u}}(t) \; . \tag{1.15}$$

Since, in general, the shape functions do not satisfy the governing differential equation, these stresses will normally not satisfy internal equilibrium at any point. The virtual work statement can be used to require that equilibrium is at least satisfied in a weak sense. An approximate solution consisting of a linear combination of n shape functions can be made to satisfy the virtual work equation for a virtual displacement space spanned by n independent virtual displacement fields. The Galerkin approach uses the same shape functions utilised to construct $\tilde{\mathbf{u}}$ to provide the n independent virtual displacement fields. In this case, the virtual work equation must be satisfied for any linear combination of the shape functions.

Denoting the virtual nodal displacements by $\delta\mathbf{u}$, this means that the virtual work equation must be satisfied for all virtual displacement fields represented by

$$\delta\mathbf{u}(\mathbf{x},t) = \mathbf{N}\delta\mathbf{u}(t) \tag{1.16}$$

and the corresponding virtual strain fields

$$\delta\boldsymbol{\varepsilon}(\mathbf{x},t) = (\mathbf{DN})\delta\mathbf{u}(t) \; . \tag{1.17}$$

Substituting (1.15) to (1.17) in (1.11), the virtual work equation becomes:

$$\delta\mathbf{u}^T(t) \left[\int_V (\mathbf{DN})^T \mathbf{E}(\mathbf{DN})\mathbf{u}(t)\, dV - \int_V \mathbf{N}^T \left(\kappa\mathbf{N}\dot{\mathbf{u}}(t) + \rho\mathbf{N}\ddot{\mathbf{u}}(t) \right) dV \right.$$

$$\left. - \int_V \mathbf{N}^T \mathbf{f}\, dV - \int_\Gamma \mathbf{N}^T \mathbf{t}\, d\Gamma \right] = 0 \; . \tag{1.18}$$

Equation (1.18) must be satisfied for any choice of $\delta\mathbf{u}(t)$, thus

$$\mathbf{K}\mathbf{u}(t) + \mathbf{C}\dot{\mathbf{u}}(t) + \mathbf{M}\ddot{\mathbf{u}}(t) = \mathbf{f} \qquad (1.19)$$

must apply, where the stiffness matrix $\mathbf{K}$, damping matrix $\mathbf{C}$, mass matrix $\mathbf{M}$ and the vector of equivalent nodal forces $\mathbf{f}$ are given by:

$$\mathbf{K} = \int_V (\mathbf{DN})^T \mathbf{E}(\mathbf{DN})\, dV , \qquad (1.20)$$

$$\mathbf{C} = \int_V \mathbf{N}^T \mathbf{N}\kappa\, dV , \qquad (1.21)$$

$$\mathbf{M} = \int_V \mathbf{N}^T \mathbf{N}\rho\, dV \quad \text{and} \qquad (1.22)$$

$$\mathbf{f} = \int_V \mathbf{N}^T \mathbf{f}\, dV + \int_\Gamma \mathbf{N}^T \mathbf{t}\, d\Gamma . \qquad (1.23)$$

In practice it is difficult, if not impossible, to determine for general finite element assemblages the element damping parameters κ, in particular because the damping properties are frequency dependent. For this reason, the matrix $\mathbf{C}$ is in general not assembled from element damping matrices, but is constructed using the mass matrix and stiffness matrix of the complete element assemblage:

$$\mathbf{C} = c_m\mathbf{M} + c_k\mathbf{K} . \qquad (1.24)$$

Experimental results should verify the amount of damping.

To solve equation (1.19) numerically, two main types of solution procedures exist, the direct integration type and the mode superposition . In the next subsection, a brief description of direct integration methods is given.

1.4 Time integration

In this subsection, numerical step-by-step procedures are explained which integrate (1.19) direct. Prior to the spatial numerical integration of finite elements, no transformation of the equation into a different form is carried out.

Direct numerical time integration is based on two ideas. First, instead of trying to satisfy (1.19) at any time t, it is aimed to satisfy this equation only at discrete time intervals Δt apart. This means that equilibrium, equivalent to static equilibrium (but including inertia and damping forces) is sought at discrete time points within the interval of solution. Therefore, all solution techniques employed in static analysis may also be used effectively in direct time integration. The second idea on, which a direct integration method is based, is that a variation of displacements, velocities, and accelerations within each time interval Δt is assumed. The form of assumption of the variation of displacements, velocities, and accelerations within each time interval determines the accuracy , stability , and computational cost of the solution procedure.

It is assumed that the displacement , velocity , and acceleration vectors at time $t_0 = 0$, denoted by $\mathbf{u}_0$, $\dot{\mathbf{u}}_0$ and $\ddot{\mathbf{u}}_0$, respectively, are known a priori, and the solution of (1.19) is required from t_0 to t_e. In the solution, the time span under consideration, t_n, is subdivided into n equal time intervals $\Delta t = \frac{t_n}{n}$.

1.4.1 Newmark algorithm

The most common time integration method for linear problems of the structure dynamics is the family of Newmark procedures, which consist of the following three equations:

$$\mathbf{f}_{n+1} = \mathbf{M}\ddot{\mathbf{u}}_{n+1} + \mathbf{C}\dot{\mathbf{u}}_{n+1} + \mathbf{K}\mathbf{u}_{n+1} , \tag{1.25}$$

$$\mathbf{u}_{n+1} = \mathbf{u}_n + \Delta t\,\dot{\mathbf{u}}_n + \frac{\Delta t^2}{2}[(1 - 2\beta)\ddot{\mathbf{u}}_{n+1} + 2\beta\ddot{\mathbf{u}}_n] , \tag{1.26}$$

$$\dot{\mathbf{u}}_{n+1} = \dot{\mathbf{u}}_n + \Delta t[(1 - \gamma)\ddot{\mathbf{u}}_n + \gamma\ddot{\mathbf{u}}_{n+1} , \tag{1.27}$$

where $\mathbf{u}_{n+1}$, $\dot{\mathbf{u}}_{n+1}$, and $\ddot{\mathbf{u}}_{n+1}$ are approximations of $\mathbf{u}$, $\dot{\mathbf{u}}$, and $\ddot{\mathbf{u}}$ for time $t_{n+1} = t_n + \Delta t$ and $\mathbf{f}_{n+1} = \mathbf{f}(t_{n+1})$.

The displacements $\mathbf{u}_n$, velocities $\dot{\mathbf{u}}_n$, and accelerations $\ddot{\mathbf{u}}_n$ of time-step n must be known. The parameters β and γ are responsible for the stability and accuracy of the method. For $\gamma = \frac{1}{2}$ the Newmark method is of second order accuracy. For $\beta = \frac{1}{4}$ and $\gamma = \frac{1}{2}$ this method becomes the trapezoid rule , or average acceleration method , which is implicit and unconditionally stable. These properties are not dependent on the chosen time-step [76]. Additionally, for linear systems the conservation of total energy is fulfilled.

For $\beta = 0$ and $\gamma = \frac{1}{2}$ the method degenerates to the Verlet algorithm , which is an explicit method, if mass- and damping matrix is a diagonal matrix. This algorithm is only conditionally stable, with the condition:

$$\Delta t \leq \Delta t_{crit} = 2/\omega_{max} , \tag{1.28}$$

where ω_{max} is the largest frequency of the system. If the condition (1.28) is not too restrictive for the considered system, this algorithm is an effective direct integration scheme [76]. For a typical two-node beam element the critical time-step is given for the Verlet algorithm by (see [14]):

$$\Delta t_{crit} = \sqrt{\frac{\rho A}{48 EI}} L^2 , \tag{1.29}$$

with EI, A, L is the stiffness, area and length of the beam, respectively.

1.4.2 Hilber-Hughes-Taylor-α method

When choosing $\gamma \neq \frac{1}{2}$ in the Newmark algorithm, an algorithmic damping of high frequencies can be achieved, but this also reduces the accuracy. For this reason, the

Hilber-Hughes-Taylor-α (HHT-α) method [75] was developed for linear problems as a variant of the Newmark method.

The HHT-α algorithm adds a third parameter α to account for the decrease with respect to accuracy that results when introducing numerical damping into the Newmark method. Setting $\alpha = 0$ reduces the problem to a standard Newmark method. Choosing $\alpha \in [-\frac{1}{3}, 0]$, $\beta = \frac{(1-\alpha)^2}{4}$ and $\gamma = \frac{1-2\alpha}{2}$ results in an unconditionally stable, second-order accurate algorithm [77], with algorithmic damping of high frequencies introduced.[1]

With these three parameters, the time-discrete equation of motion in the HHT-α method is written as

$$\mathbf{f}_{n+1+\alpha} = \mathbf{M}\ddot{\mathbf{u}}_{i+1} + (1+\alpha)\mathbf{C}\dot{\mathbf{u}}_{i+1} - \alpha\mathbf{C}\dot{\mathbf{u}}_i + (1+\alpha)\mathbf{K}\mathbf{u}_{i+1} - \alpha\mathbf{K}\mathbf{u}_i \,. \tag{1.30}$$

with

$$\mathbf{f}_{n+1+\alpha} = \mathbf{f}(t_{n+1} + \alpha\Delta t) \,. \tag{1.31}$$

The standard Newmark finite difference formulas are used as approximations for $\mathbf{u}_{i+1}$ and $\dot{\mathbf{u}}_{i+1}$,

$$\mathbf{u}_{i+1} = \tilde{\mathbf{u}}_{i+1} + \beta\Delta t^2 \ddot{\mathbf{u}}_{i+1} \,, \tag{1.32}$$

$$\dot{\mathbf{u}}_{i+1} = \tilde{\dot{\mathbf{u}}}_{i+1} + \gamma\Delta t \ddot{\mathbf{u}}_{i+1} \,, \tag{1.33}$$

where the predictor variables, $\tilde{\mathbf{u}}_{i+1}$ and $\tilde{\dot{\mathbf{u}}}_{i+1}$ are defined as

$$\tilde{\mathbf{u}}_{i+1} = \mathbf{u}_i + \Delta t \dot{\mathbf{u}}_i + \Delta t^2 \left(\frac{1}{2} - \beta \right) \ddot{\mathbf{u}}_i \,, \tag{1.34}$$

$$\tilde{\dot{\mathbf{u}}}_{i+1} = \dot{\mathbf{u}}_i + (1-\gamma)\Delta t \ddot{\mathbf{u}}_i \,. \tag{1.35}$$

An implicit update equation can be achieved with $\mathbf{u}_{i+1}$ as the only unknown by rearranging (1.32) and substituting the result along with (1.33) into the equation of motion (1.30). If the left hand coefficient matrix is denoted as $\tilde{\mathbf{L}}$ and the right hand contributions excluding the force vector as $\tilde{\mathbf{R}}$ then the update equation is

$$\tilde{\mathbf{L}}\mathbf{u}_{i+1} = \tilde{\mathbf{R}}_{i+1} + \mathbf{f}_{i+1+\alpha} \,, \tag{1.36}$$

where

$$\tilde{\mathbf{L}} = \mathbf{M} + \gamma\Delta t\mathbf{C} + (1+\alpha)\beta\Delta t^2\mathbf{K} \,, \tag{1.37}$$

and

$$\tilde{\mathbf{R}}_{i+1} = [\mathbf{M} + \gamma\Delta t\mathbf{C}]\tilde{\mathbf{u}}_{i+1} - (1+\alpha)\beta\Delta t^2\mathbf{C}\tilde{\dot{\mathbf{u}}}_{i+1} + \alpha\beta\Delta t^2 [\mathbf{C}\dot{\mathbf{u}}_i + \mathbf{K}\mathbf{u}_i] \,. \tag{1.38}$$

The HHT-α integration scheme is used in the coupled FE/SBFE model of this work.

---

[1] For numerical simulations in this work $\alpha = -\frac{1}{2}$; $\beta = \frac{25}{64}$; $\gamma = \frac{3}{4}$ is chosen.

2. Boundary element method

In order to check the accuracy of the presented coupled finite element/scaled boundary finite element (FE/SBFE) procedure, the boundary element method (BEM) is used to compute comparison solutions. In the following, a short overview of the BEM utilised for elastodynamic problems in the time-domain will be given. Detailed derivations and application examples can be found, e.g., in Beer [15], Gaul et al. [55] or Wrobel et al. [149].

2.1 Integral formulation for elastodynamics

The most commonly used possibility to obtain an integral formulation is either a formal mathematical way by using weighted residuals or a more physical one by using the reciprocal work theorem . In this subsection, the weighted residual statement will be used.

The weighted residual statement for elastodynamics is achieved by setting the inner product of the dynamic equilibrium (1.7), in indicial notation:

$$\sigma_{ij,j} + f_i - \kappa \dot{u}_i - \rho \ddot{u}_i = 0 \,, \tag{2.1}$$

and the fundamental solution for the displacements U_{ij}^* to zero. The fundamental solution is formulated in the spatial variable and the convolution is performed with respect to the time variable.

By neglecting body forces f_i, damping $\kappa \dot{u}_i$ and omitting the spatial dependency in the notation:

$$\sigma_{ij}(t) = \sigma_{ij}(\mathbf{x},t) \,,$$
$$\ddot{u}_i(t) = \ddot{u}_i(\mathbf{x},t) \,,$$
$$U_{ij}^*(t-\tau) = U_{ij}^*(\mathbf{x},\boldsymbol{\xi},t-\tau) \,,$$

the weighted residual statement can be formulated:

$$\int_0^t \int_\Omega \left[\sigma_{ik,k}(\tau) - \rho \ddot{u}_i(\tau)\right] U_{ij}^*(t-\tau) \, d\Omega \, d\tau = 0 \,, \tag{2.2}$$

where the simplified dynamic equilibrium

$$\sigma_{ik,k}(t) - \rho\ddot{u}_i(t) = 0 \tag{2.3}$$

is used. Equation (2.2) states that the error introduced to solve the differential equation (2.3) is zero averaged over the domain Ω and time t. To choose the fundamental solution U_{ij}^* as the weighting function is at this point arbitrary, i.e., every smooth function could be used. But, in order to formulate a boundary element equation, the weighting function must be chosen as fundamental solution.

The constitutive equation (1.4) is valid for an anisotropic homogeneous elastic medium, however, due to the lack of fundamental solutions for the general anisotropic case, the following is restricted to an isotropic homogeneous continuum.

Next, the two parts in the brackets $[\dots]$ of (2.2) are treated separately. Partial integration of the first term using the divergence theorem and the Cauchy theorem $\mathbf{t} = \boldsymbol{\sigma}\mathbf{n}$ (in indicial notation: $t_i = \sigma_{ij}n_j$) yields

$$\int_{\Omega} \sigma_{ik,k}(t) * U_{ij}^*(t) \, d\Omega$$

$$= \int_{\Gamma} \sigma_{ik}(t)n_k * U_{ij}^*(t) \, d\Gamma - \int_{\Omega} \sigma_{ik}(t) * U_{ij,k}^*(t) \, d\Omega$$

$$= \int_{\Gamma} t_i(t) * U_{ij}^*(t) \, d\Gamma - \int_{\Omega} \sigma_{ik}(t) * U_{ij,k}^*(t) \, d\Omega \,, \tag{2.4}$$

where the symbol "$*$" denotes the time convolution product of two functions of time, defined as:

$$a(t) * b(t) = \int_0^t a(\tau)b(t - \tau) \, d\tau \quad (t \geq 0) \,. \tag{2.5}$$

A second partial integration is applied to the last integral in (2.4):

$$\int_{\Omega} \sigma_{ik}(t) * U_{ij,k}^*(t) \, d\Omega$$

$$= \int_{\Omega} C_{ikmn}u_{m,n}(t) * U_{ij,k}^*(t) \, d\Omega$$

$$= \int_{\Gamma} u_m(t) * C_{ikmn}U_{ij,k}^*(t)n_n \, d\Gamma - \int_{\Omega} u_m(t) * C_{ikmn}U_{ij,kn}^*(t) \, d\Omega$$

$$= \int_{\Gamma} u_i(t) * T_{ij}^*(t) \, d\Gamma - \int_{\Omega} u_i(t) * S_{ijk,k}^*(t) \, d\Omega \tag{2.6}$$

using the symmetric properties of the material tensor C_{ijkl} for isotropic material and the linear geometry relation (1.1):

$$\sigma_{ij} = C_{ijkl}\frac{1}{2}\left(u_{k,l} + u_{l,k}\right) = \frac{1}{2}\left(C_{ijkl}u_{k,l} + C_{ijlk}u_{l,k}\right) = C_{ijkl}u_{k,l} \,. \tag{2.7}$$

In (2.6), S_{ijk}^* denotes the fundamental solution for the stress tensor calculated by the constitutive equation (2.7) with the fundamental solutions for the displacements and T_{ij}^* is the traction fundamental solution . These manipulations shift the divergence operator from the stress tensor $\boldsymbol{\sigma}$ to the fundamental solution.

The inertia term in the weighted residual statement (2.2) is also mapped on the fundamental solution U_{ij}^* using two partial integrations with respect to time:

$$\ddot{u}_i(t) * U_{ij}^*(t)$$

$$= \int_0^t \frac{\partial^2}{\partial \tau^2} u_i(\tau) U_{ij}^*(t-\tau)\, d\tau$$

$$= \left[\frac{\partial}{\partial \tau} u_i(\tau) U_{ij}^*(t-\tau) - u_i(\tau)\frac{\partial}{\partial \tau} U_{ij}^*(t-\tau) \right]_0^t + \int_0^t u_i(\tau)\frac{\partial^2}{\partial \tau^2} U_{ij}^*(t-\tau)\, d\tau. \qquad (2.8)$$

With the assumption of vanishing initial conditions for the displacements and vanishing initial fundamental solutions

$$u_i(t=0) = \dot{u}_i(t=0) = U_{ij}^*(t-\tau=0) = \dot{U}_{ij}^*(t-\tau=0) = 0 \qquad (2.9)$$

the integral (2.2) reads

$$\int_\Omega u_i(t) * (S_{ijk,k}^*(t) - \rho \ddot{U}_{ij}^*(t))\, d\Omega$$

$$+ \int_\Gamma t_i(t) * U_{ij}^*(t)\, d\Gamma - \int_\Gamma u_i(t) * T_{ij}^*(t)\, d\Gamma = 0 . \qquad (2.10)$$

The fundamental solutions[1] have to fulfil the inhomogeneous differential equation

$$S_{ijk,k}^*(\mathbf{x},\boldsymbol{\xi},t-\tau) - \rho \ddot{U}_{ij}^*(\mathbf{x},\boldsymbol{\xi},t-\tau) = -\delta(t-\tau)(\mathbf{x}-\boldsymbol{\xi}) . \qquad (2.11)$$

Using condition (2.11) for determining the fundamental solutions and using the properties of the Dirac distribution δ, the *boundary integral equation* is achieved

$$u_j(\boldsymbol{\xi},t) = \int_\Gamma t_i(t) * U_{ij}^*(t)\, d\Gamma - \int_\Gamma u_i(t) * T_{ij}^*(t)\, d\Gamma \qquad (2.12)$$

with the boundary displacement u_i and traction t_i as the only unknowns.

Boundary extension around load point

At every point $\mathbf{x} \in \Gamma$ either displacements or tractions must be prescribed for a well posed problem. To determine the missing boundary data $\boldsymbol{\xi}$ has to be moved onto the

---

[1] A fundamental solution has to be interpreted in the sense of a distribution. It is the response due to an impulse loading (point load) at the point $\boldsymbol{\xi}$ of a system governed by a differential operator.

boundary Γ. To assure that $\boldsymbol{\xi}$ remains inside the domain, the boundary is augmented by a small extension in the vicinity of the load point. While in principle the shape of this extension can be arbitrary, it is advantageous to use a symmetric extension, since this relates to the concept of the Cauchy principle value of the integral. Therefore, the boundary is extended by part of a sphere Γ_ε with radius $\varepsilon = |x_i - \xi_i|$, whose centre coincidences with the load point as shown in Fig. 2.1.

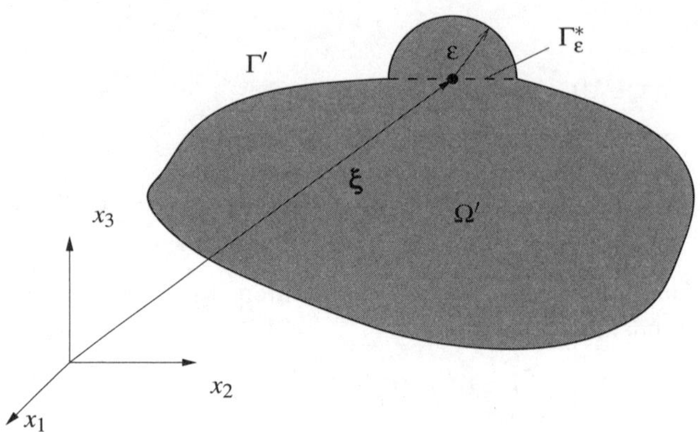

Fig. 2.1. Cross section through domain with spherical extension around load point

By taking the limit $\varepsilon \to 0$, the modified boundary Γ' approaches the original boundary $\Gamma = \lim_{\varepsilon \to 0} \Gamma'$ with $\Gamma' = \Gamma - \Gamma_\varepsilon^* + \Gamma_\varepsilon$, and the load point becomes a boundary point. With this boundary extension, (2.12) can be written as:

$$u_j(\boldsymbol{\xi},t) = \lim_{\varepsilon \to 0} \int\limits_{\Gamma - \Gamma_\varepsilon^*} t_i(t) * U_{ij}^*(t)\, d\Gamma - \lim_{\varepsilon \to 0} \int\limits_{\Gamma - \Gamma_\varepsilon^*} u_i(t) * T_{ij}^*(t)\, d\Gamma$$

$$+ \lim_{\varepsilon \to 0} \int\limits_{\Gamma_\varepsilon} t_i(t) * U_{ij}^*(t)\, d\Gamma - \lim_{\varepsilon \to 0} \int\limits_{\Gamma_\varepsilon} u_i(t) * T_{ij}^*(t)\, d\Gamma \,. \qquad (2.13)$$

Integration over boundary $\Gamma - \Gamma_\varepsilon^*$

The time-dependent fundamental solutions behave like the elastostatic ones in the limit $\boldsymbol{\xi} \to \mathbf{x}$ [116, 55]. Thus, it is sufficient to investigate in the following only the singular behaviour of the static fundamental solution.

When approaching the load point ($r = \varepsilon \to 0$), for the static displacement and traction fundamental solution it is obtained in the three-dimensional domain:

$$U_{ij}^* \sim \frac{1}{r} \quad \text{for} \quad r \to 0 \,, \text{ and} \qquad (2.14)$$

$$T_{ij}^* \sim \frac{1}{r^2} \quad \text{for} \quad r \to 0 \,. \qquad (2.15)$$

Obviously, the Dirichlet fundamental solution U_{ij}^* is weakly singular, while the Neumann fundamental solution T_{ij}^* is strongly singular.

The weakly singular integral exists as an improper integral, i.e., despite the fact that $U_{ij}^* \to \infty$ for $r \to 0$, the integral over U_{ij}^* is bounded at $r = 0$. Thus, the first part on the right hand side of (2.13) can be regularised by a coordinate transformation for $t_i(t) * U_{ij}^*(t)$.

In contrast, the integral over the strongly singular kernel T_{ij}^* is unbounded. However, if the boundary extension Γ_ε is contracted symmetrically towards the load point, the integral remains finite in the limit $\varepsilon \to 0$. This can be achieved, for example, by using a sphere with radius $r = \varepsilon$ (see above). If the load point is approached with an arbitrary, non-symmetric surface, the second term on the right hand side of (2.13) becomes unbounded. However, the sum of both remains finite and fulfils (2.12). This integral has to be defined in the sense of a Cauchy principle value for $u_i(t) * T_{ij}^*(t)$.

Integration over boundary extension Γ_ε

The displacement and stress field are continuous at the singularity, so that

$$\lim_{r \to 0} \int_{\Gamma_\varepsilon} t_j(\mathbf{x}) U_{ij}^*(\mathbf{x}, \boldsymbol{\xi}) \, d\Gamma_\varepsilon = \sigma_{jk}(\boldsymbol{\xi}) \lim_{r \to 0} \int_{\Gamma_\varepsilon} U_{ij}^*(\mathbf{x}, \boldsymbol{\xi}) n_k(\mathbf{x}) \, d\Gamma_\varepsilon , \qquad (2.16)$$

$$\lim_{r \to 0} \int_{\Gamma_\varepsilon} u_j(\mathbf{x}) T_{ij}^*(\mathbf{x}, \boldsymbol{\xi}) \, d\Gamma_\varepsilon = u_j(\boldsymbol{\xi}) \lim_{r \to 0} \int_{\Gamma_\varepsilon} T_{ij}^*(\mathbf{x}, \boldsymbol{\xi}) \, d\Gamma_\varepsilon . \qquad (2.17)$$

The integration over Γ_ε is carried out in spherical coordinates (radius r, polar angle Θ_1 and azimuth angle Θ_2, see Fig. 2.2) with origin at $\boldsymbol{\xi}$.

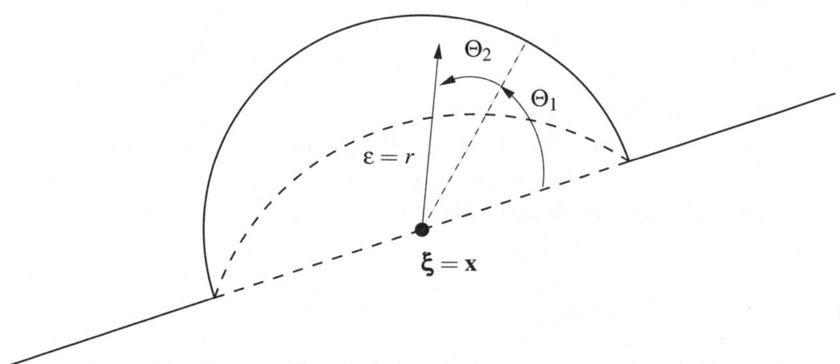

Fig. 2.2. Spherical boundary extension around load point with polar coordinates (r, Θ_1, Θ_2)

The coordinates of the radius r_i are given by:

$$r_i = x_i - \xi_i = \begin{bmatrix} \sin\Theta_1 \cos\Theta_2 \\ \sin\Theta_1 \sin\Theta_2 \\ \cos\Theta_1 \end{bmatrix} r , \qquad (2.18)$$

the radius $r = \sqrt{r_i r_i}$ and $r_{,i} = \frac{r_i}{r}$.

The outward normal vector of the sphere n_i is given by:

$$n_i = \begin{bmatrix} \sin\Theta_1\cos\Theta_2 \\ \sin\Theta_1\sin\Theta_2 \\ \cos\Theta_1 \end{bmatrix} , \tag{2.19}$$

thus, $n_i = r_{,i}$ and $n_i n_i = r_{,i} r_{,i} = 1$. The fundamental solutions can be expressed in spherical coordinates, see Gaul [55]:

$$U_{ij}^* = \frac{1}{r} G_{ij}^u(\Theta_1,\Theta_2) , \tag{2.20}$$

$$T_{ij}^* = \frac{1}{r^2} G_{ij}^t(\Theta_1,\Theta_2) . \tag{2.21}$$

With the element area $d\Gamma$ computed with $d\Gamma_\varepsilon = r^2\sin\Theta_1\,d\Theta_1\,d\Theta_2$, integration of (2.16) and (2.17) leads to:

$$\lim_{r\to 0}\int_{\Gamma_\varepsilon} t_j(\mathbf{x})U_{ij}^*(\mathbf{x},\boldsymbol{\xi})\,d\Gamma = \sigma_{jk}(\boldsymbol{\xi})\lim_{r\to 0}\int_{\Theta_2}\int_{\Theta_1}\frac{1}{r}G_{ij}^u(\Theta_1,\Theta_2)n_k r^2\sin\Theta_1\,d\Theta_1\,d\Theta_2 = 0 , \tag{2.22}$$

the weakly singular integral is zero over Γ_ε, and

$$\lim_{r\to 0}\int_{\Gamma_\varepsilon} T_{ij}^*\,d\Gamma = \int_{\Theta_2}\int_{\Theta_1} G_{ij}^t(\Theta_1,\Theta_2)\sin\Theta_1\,d\Theta_1\,d\Theta_2 =: c_{ij} , \tag{2.23}$$

the strong singular integral yields a finite value (c_{ij}).

For a smooth boundary ($0\le\Theta_1<\pi/2$ and $0\le\Theta_2\le 2\pi$)

$$\int_{\Theta_2}\int_{\Theta_1} G_{ij}^t(\Theta_1,\Theta_2)\sin\Theta_1\,d\Theta_1\,d\Theta_2 = -\frac{1}{2}\delta_{ij} \tag{2.24}$$

is obtained.

The same splitting of the fundamental solutions in a singular elastostatic part and a regular dynamic part is also possible for two-dimensional solutions. For arbitrary boundary points, e.g., corners, the term c_{ij} is determined following a procedure given in [97].

Boundary integral equation

Finally, these considerations result in the boundary integral equation

$$c_{ij}(\boldsymbol{\xi})u_i(\boldsymbol{\xi},t)$$

$$= \int_\Gamma t_i(t) * U_{ij}^*(t)\,d\Gamma - \oint_\Gamma u_i(t) * T_{ij}^*(t)\,d\Gamma$$

$$= \int_0^t\int_\Gamma t_i(\mathbf{x},\tau)U_{ij}^*(\mathbf{x},\boldsymbol{\xi},t-\tau)\,d\Gamma\,d\tau - \int_0^t\oint_\Gamma u_i(\mathbf{x},\tau)T_{ij}^*(\mathbf{x},\boldsymbol{\xi},t-\tau)\,d\Gamma\,d\tau . \tag{2.25}$$

For an arbitrary domain the integral equation can not be solved analytically. Therefore, a discretisation is necessary for a solution procedure, leading to the boundary element formulation (see next subsection).

2.2 Numerical implementation

The variables $\mathbf{u}, \mathbf{t}$ over the boundary Γ are represented using interpolation techniques initially developed for the FEM. The boundary surface Γ is discretised by E elements where polynomial spatial shape functions $N_e^f(\boldsymbol{\eta})$ with F nodes are defined (see, e.g., the books of Beer [15], Gaul [55], Bonnet [26] or Paris et al. [108]). Hence, with the time-dependent nodal values $\mathbf{u}^f(t)$ and $\mathbf{t}^f(t)$ the displacements and tractions are approximated, respectively, by

$$\tilde{\mathbf{u}}(\mathbf{x},t) = \sum_{f=1}^{F} N^f(\boldsymbol{\eta})\,\mathbf{u}^f(t)\,, \tag{2.26}$$

$$\tilde{\mathbf{t}}(\mathbf{x},t) = \sum_{f=1}^{F} N^f(\boldsymbol{\eta})\,\mathbf{t}^f(t)\,, \tag{2.27}$$

where $\mathbf{x}$ and $\boldsymbol{\eta}$ are related by the mapping (1.13). Inserting these shape functions in the integral equation (2.25) and assembling the whole system with E elements yields

$$c_{ij}(\boldsymbol{\xi})\tilde{u}_i(\boldsymbol{\xi},t)$$

$$= \sum_{e=1}^{E} \sum_{f=1}^{F} \left[\int_{\Gamma_e} U_{ij}^*(\mathbf{x},\boldsymbol{\xi},t) N_e^f(\boldsymbol{\eta})\,d\Gamma_e * t_i^{ef}(t) - \oint_{\Gamma_e} T_{ij}^*(\mathbf{x},\boldsymbol{\xi},t) N_e^f(\boldsymbol{\eta})\,d\Gamma_e * u_i^{ef}(t) \right]\,. \tag{2.28}$$

If body forces should be considered, the integral I_b

$$I_b = \int_{\Omega} b_i(t) * U_{ij}^*(t)\,d\Omega \tag{2.29}$$

has to be evaluated. Here, no test functions are necessary because the integrand is a priori known and needs only to be integrated numerically over the domain. In some special cases, e.g., for the gravity force, the domain integral can be shifted on the boundary [9].

The next step is the time discretisation. Here, interpolation functions for the time variable are introduced as well, and subsequent the time integration in each time-step is performed analytically. This method was first introduced by Mansur [95].

A different time-stepping procedure – based on the convolution quadrature – could also be utilised, see Schanz [116].

2.3 Time domain boundary element formulation

The time-stepping procedure proposed by Mansur [95] and later extended to non-zero initial conditions by Antes [6] approximates the spatial behaviour with polynomial shape functions . Additionally, the time t is approximated by polynomials as well, mostly linear for displacement and constant for traction after discretising t in N equal time-steps Δt. A different approach is proposed by Karabalis and Rizos [82] where spline functions are used to approximate the time history.

Using linear polynomials for the displacements and constant ones for the tractions, the test functions are

$$\tilde{u}_i(\mathbf{x},\tau) = \sum_{e=1}^{E}\sum_{f=1}^{F}\sum_{m=1}^{N} N_e^f(\boldsymbol{\eta})\left(U_{if}^{em-1}\frac{t_m-\tau}{\Delta t}+U_{if}^{em}\frac{\tau-t_{m-1}}{\Delta t}\right)t_i(\tau,\mathbf{x})$$

$$= \sum_{e=1}^{E}\sum_{f=1}^{F}\sum_{m=1}^{N} N_e^f(\boldsymbol{\eta})T_{if}^{em}\,, \tag{2.30}$$

with nodal values U_{if}^{em} and T_{if}^{em} defined at the time and spatial collocation points.[2]

Inserting these test functions in the integral equation (2.25) yields

$$c_{ij}(\boldsymbol{\xi})\tilde{u}_i(\boldsymbol{\xi},t)$$

$$= \sum_{e=1}^{E}\sum_{f=1}^{F}\sum_{m=1}^{N}\left[\int_{\Gamma_e} N_e^f(\boldsymbol{\eta})\int_{t_{m-1}}^{t_m} U_{ij}(\mathbf{x},\boldsymbol{\xi},t-\tau)\,d\tau\,d\Gamma_e\right.$$

$$\left. -\oint_{\Gamma_e} N_e^f(\boldsymbol{\eta})\int_{t_{m-1}}^{t_m} T_{ij}(\mathbf{x},\boldsymbol{\xi},t-\tau)\left(U_{if}^{em-1}\frac{t_m-\tau}{\Delta t}+U_{if}^{em}\frac{\tau-t_{m-1}}{\Delta t}\right)d\tau d\Gamma_e\right]\,. \tag{2.31}$$

Due to the properties of the time-dependent fundamental solutions – they consist of Dirac distributions and Heaviside functions – the time integration within each time-step is performed analytically. For the spatial integration Gaussian quadrature formulas are used except when $\mathbf{x}$ coincides $\boldsymbol{\xi}$. In this singular case, the same regularisation methods as described above are applied. Point collocation and the properties of the fundamental solutions – causality and independence on t but dependence on the difference $(t-\tau)$ – yield a time-stepping procedure

$$\mathbf{C}^1\mathbf{d}^n = \mathbf{D}^1\bar{\mathbf{d}}^n + \sum_{k=2}^{\bar{n}}\left(\mathbf{U}^k\mathbf{t}^{n-k+1}-\mathbf{T}^k\mathbf{u}^{n-k+1}\right)\quad n=1,2,\ldots,N \tag{2.32}$$

with the matrix of the time-integrated fundamental solutions of the tractions $\mathbf{T}^k$ and displacements $\mathbf{U}^k$. In the matrices $\mathbf{C}^1$ and $\mathbf{D}^1$ are the time-integrated fundamental

---

[2] Other methods exist, where it is attempted to reduce the error over a *region*. However, the use of weighted residual methods, of which the Galerkin method is a special case, does not result in the expected increase in accuracy [15, 18].

solutions of the first time-step assembled concerning the unknown boundary data $\mathbf{d}$ and the known boundary data $\bar{\mathbf{d}}$, respectively. The upper limit of the sum $\bar{n}$ is determined by the fact that for three-dimensional analysis the fundamental solutions are zero after passing of the shear wave, i.e., $\bar{n} = \min(n, \frac{r_{max}}{c_2 \Delta t} + 2)$, and for a two-dimensional analysis $\bar{n} = n$ applies. For more details about this direct approach in time domain see, e.g., [44] or [96].

3. Scaled boundary finite element method

In the following, the main idea and derivation of the scaled boundary finite element equation for elastodynamics in the time domain is presented.

Wolf and Song have developed the so-called consistent infinitesimal finite element cell method [147] for wave propagation in unbounded media, including incompressible media subject to diffusion, the vector wave equation or the scalar wave equation [123, 124, 125, 126]. This method is nowadays termed the scaled boundary finite element method (SBFEM) [148]. Main application of the SBFEM is the analysis of wave propagation problems in an elastic half-space, which occur, e.g., in soil-structure interaction simulations.

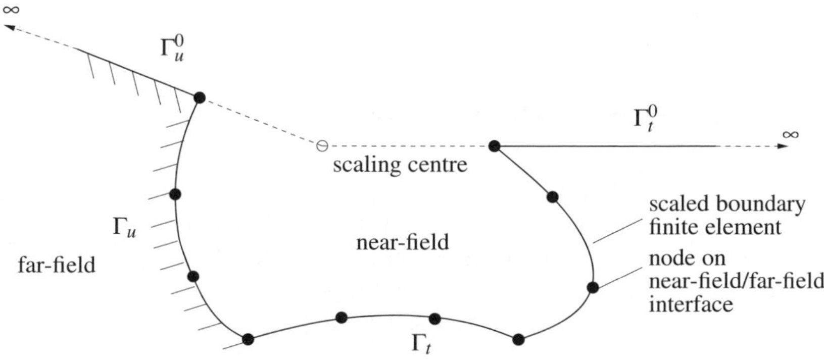

Γ: boundary (near-field/far-field interface)
Γ^0: part of boundary, which has not to be discretised
Γ_u: boundary with prescribed displacements
Γ_t: boundary with prescribed tractions

Fig. 3.1. SBFE discretisation of unbounded media

The SBFEM was mainly developed for the discretisation of an infinite domain which acts as absorbing boundary condition. For the simulation of soil-structure interaction problems , the domain is divided into a so-called near-field and far-field part , of which the far-field represents the infinite space, see Fig. 3.1. To avoid wave reflections at this artificially introduced boundary, the near-field/far-field interface is discretised with scaled boundary finite elements. To establish an absorbing boundary condition, the SBFEs are only necessary at this interface, the free surface (marked with Γ^0 in Fig. 3.1) has not to be discretised. The dimension of the SBFE discretisation is – as in boundary element calculations – reduced by one, compared to the dimension of the considered problem.

Fig. 3.2 shows a typical discretisation of a soil-structure interaction problem . The near-field is discretised using finite elements and may include the structure (building, dam, or similar) and a portion of the surrounding soil. Within this part of the domain non-linear effects can be taken into account. The near-field/far-field interface is marked with Γ^t. At this interface, due to the direct FE/SBFE coupling, the nodes of both element types have to be congruent.

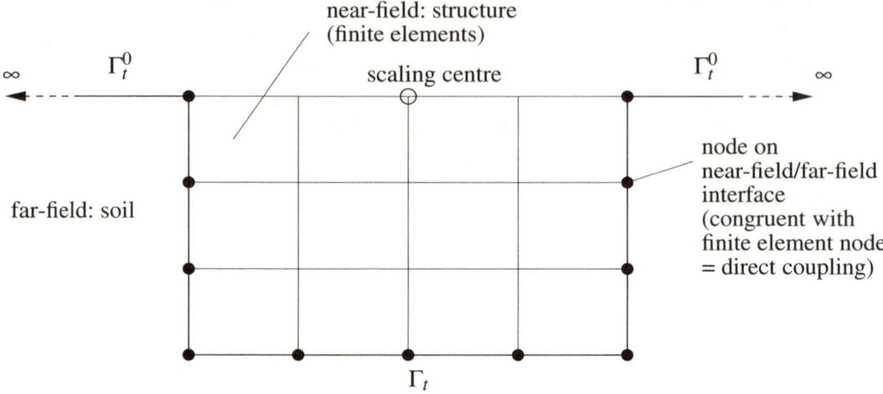

Fig. 3.2. Standard application: soil-structure interaction

An advantage of using the SBFEM is, besides others, that the calculated influence of the infinite far-field can be stored in form of acceleration unit-impulse influence matrices in the time domain. Therefore, the SBFE calculation can be performed before starting the simulation of the whole problem and must not be repeated if different load cases or geometries of the near-field have to be investigated.

3.1 Scaled boundary transformation

With the SBFEM, the geometry of a three-dimensional problem is described by two-dimensional finite elements with local coordinates η, ζ on the boundary and a radial coordinate ξ. This scaled boundary coordinate system is related to the Cartesian coordinate system (x_i) by the so-called scaled boundary transformation, which describes similarity. The coordinate ξ measures the distance from the scaling centre O, being defined as $\xi = 1$ when crossing the boundary (see Fig. 3.3). Hence, ξ contains a scaling factor.

The scaled boundary coordinate system permits a numerical treatment in the circumferential directions based on a *weighted residual technique* , as in the theory of

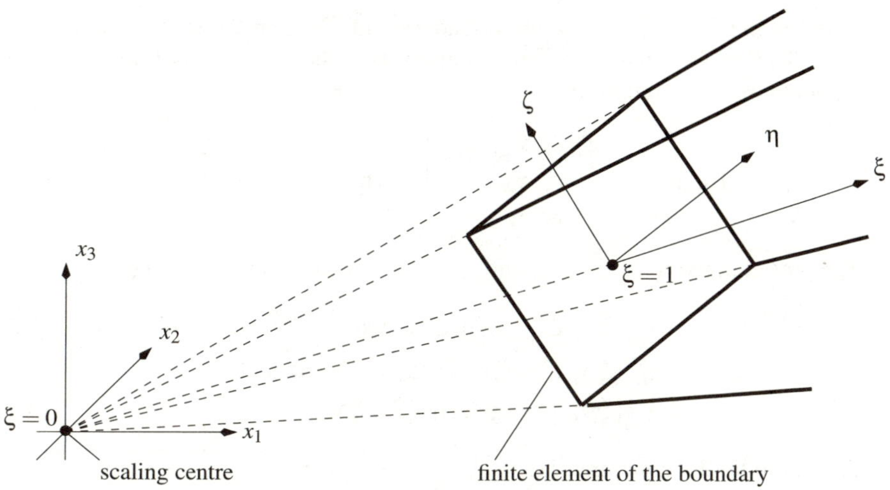

Fig. 3.3. Scaled boundary transformation for three-dimensional problems: finite element and coordinate systems

finite elements. Therewith, the partial differential equations will be transformed into ordinary differential equations in the radial coordinate ξ, where their coefficients are determined by the finite element approximation in the circumferential directions.

For an unbounded medium, the radial coordinate ξ points from the boundary towards infinity (see Fig. 3.3) where the boundary condition at infinity (radiation condition) can be incorporated in the analytical solution.

3.2 Governing equations in scaled boundary coordinates

The dynamic equilibrium (1.7) reads in the frequency domain:

$$\mathbf{D}^{T}\hat{\boldsymbol{\sigma}}+\hat{\mathbf{f}}+\omega^{2}\rho\hat{\mathbf{u}}=\mathbf{0}\,,\tag{3.1}$$

where $\hat{\boldsymbol{\sigma}}$, $\hat{\mathbf{f}}$ and $\hat{\mathbf{u}}$ stand for spatial-dependent amplitudes of stresses , body forces , and displacements , respectively[1], ρ denotes the material density .

The geometry of a finite element on the boundary is represented by interpolating its nodal coordinates $\mathbf{x}_i$ using the local coordinates η and ζ:

$$x_{i}(\eta,\zeta)=\mathbf{N}(\eta,\zeta)\mathbf{x}_{i}\,,\tag{3.2}$$

with the mapping functions $\mathbf{N}=\mathbf{N}(\eta\zeta)=[N_{1}\,N_{2}\,\ldots]$. A point in the domain $\mathbf{x}(\xi,\eta,\zeta)$ is obtained by scaling the corresponding point $\mathbf{x}(\eta,\zeta)$ on the boundary:

$$\mathbf{x}(\xi,\eta,\zeta)=\xi\mathbf{x}(\eta,\zeta)\,,\tag{3.3}$$

---

[1] For abbreviation the hat (ˆ) is omitted for amplitudes in the following.

with the scaling factor $\xi = 1$ on the boundary and $\xi = 0$ in the scaling centre (see Fig. 3.3). To transform the differential operator $\mathbf{D}$ to the (ξ, η, ζ)–coordinate system, the Jacobian matrix is required:

$$\mathbf{J}(\xi, \eta, \zeta) = \begin{bmatrix} x_{1,\xi} & x_{2,\xi} & x_{3,\xi} \\ x_{1,\eta} & x_{2,\eta} & x_{3,\eta} \\ x_{1,\zeta} & x_{2,\zeta} & x_{3,\zeta} \end{bmatrix} \quad , \text{with } x_i = x_i(\xi, \eta, \zeta) . \tag{3.4}$$

The partial derivatives $x_{i,\xi}$, $x_{i,\eta}$ and $x_{i,\zeta}$ are calculated via (3.2) and (3.3):

$$x_{i,\xi}(\xi, \eta, \zeta) = x_i(\eta, \zeta) = \mathbf{N}x_i , \tag{3.5}$$

$$x_{i,\eta}(\xi, \eta, \zeta) = \xi x_{i,\eta}(\eta, \zeta) = \xi \mathbf{N}_{,\eta} x_i \text{ and} \tag{3.6}$$

$$x_{i,\zeta}(\xi, \eta, \zeta) = \xi x_{i,\zeta}(\eta, \zeta) = \xi \mathbf{N}_{,\zeta} x_i \tag{3.7}$$

leading to the Jacobian matrix:

$$\mathbf{J}(\xi, \eta, \zeta) = \begin{bmatrix} x_1 & x_2 & x_3 \\ \xi x_{1,\eta} & \xi x_{2,\eta} & \xi x_{3,\eta} \\ \xi x_{1,\zeta} & \xi x_{2,\zeta} & \xi x_{3,\zeta} \end{bmatrix} = \begin{bmatrix} 1 & & \\ & \xi & \\ & & \xi \end{bmatrix} \mathbf{J}(\eta, \zeta) \tag{3.8}$$

$$\text{with } \mathbf{J}(\eta, \zeta) = \begin{bmatrix} x_1 & x_2 & x_3 \\ x_{1,\eta} & x_{2,\eta} & x_{3,\eta} \\ x_{1,\zeta} & x_{2,\zeta} & x_{3,\zeta} \end{bmatrix} \tag{3.9}$$

$$\text{and } x_i = x_i(\eta, \zeta) .$$

To calculate the determinant of $\mathbf{J}$, the base vectors $\mathbf{x}(\eta, \zeta)$, $\mathbf{x}_{,\eta}(\eta, \zeta)$ and $\mathbf{x}_{,\zeta}(\eta, \zeta)$ of the transformed coordinate system are introduced (see Fig. 3.4):

$$\mathbf{x}(\eta, \zeta) = \begin{bmatrix} x_1 \\ x_2 \\ x_3 \end{bmatrix} , \tag{3.10}$$

$$\mathbf{x}_{,\eta}(\eta, \zeta) = \begin{bmatrix} x_{1,\eta} \\ x_{2,\eta} \\ x_{3,\eta} \end{bmatrix} \quad \text{and} \tag{3.11}$$

$$\mathbf{x}_{,\zeta}(\eta, \zeta) = \begin{bmatrix} x_{1,\zeta} \\ x_{2,\zeta} \\ x_{3,\zeta} \end{bmatrix} \tag{3.12}$$

$$\text{with } x_i = x_i(\eta, \zeta) ,$$

thus, the determinant of $\mathbf{J}$ can be calculated by:

$$|\mathbf{J}| = \mathbf{x}(\eta, \zeta) \cdot (\mathbf{x}_{,\eta}(\eta, \zeta) \times \mathbf{x}_{,\zeta}(\eta, \zeta)) , \tag{3.13}$$

and the infinitesimal volume dV for any ξ is:

$$dV = \mathbf{x}_{,\xi}(\xi, \eta, \zeta) \cdot \left[\mathbf{x}_{,\eta}(\xi, \eta, \zeta) \times \mathbf{x}_{,\zeta}(\xi, \eta, \zeta) \right] d\xi d\eta d\zeta . \tag{3.14}$$

With $\mathbf{x}(\xi,\eta,\zeta) = \xi\mathbf{x}(\eta,\zeta)$ and by means of (3.13) the infinitesimal volume is determined by:

$$dV = \xi^2 |\mathbf{J}|\, d\xi\, d\eta\, d\zeta \,. \tag{3.15}$$

The derivatives with respect to x_i can be expressed as a function of the scaled boundary coordinates ξ,η,ζ with (3.8) by:

$$\begin{bmatrix} \frac{\partial}{\partial x_1} \\ \frac{\partial}{\partial x_2} \\ \frac{\partial}{\partial x_3} \end{bmatrix} = \mathbf{J}^{-1}(\xi,\eta,\zeta) \begin{bmatrix} \frac{\partial}{\partial \xi} \\ \frac{\partial}{\partial \eta} \\ \frac{\partial}{\partial \zeta} \end{bmatrix} = \mathbf{J}^{-1}(\eta,\zeta) \begin{bmatrix} \frac{\partial}{\partial \xi} \\ \frac{1}{\xi}\frac{\partial}{\partial \eta} \\ \frac{1}{\xi}\frac{\partial}{\partial \zeta} \end{bmatrix} \,. \tag{3.16}$$

The inverse of Jacobian $\mathbf{J}^{-1}(\eta,\zeta)$ is given by:

$$\mathbf{J}^{-1}(\eta,\zeta) = \frac{1}{|\mathbf{J}|} \begin{bmatrix} x_{2,\eta}x_{3,\zeta} - x_{3,\eta}x_{2,\zeta} & x_3 x_{2,\zeta} - x_2 x_{3,\zeta} & x_2 x_{3,\eta} - x_3 x_{2,\eta} \\ x_{3,\eta}x_{1,\zeta} - x_{1,\eta}x_{3,\zeta} & x_1 x_{3,\zeta} - x_3 x_{1,\zeta} & x_3 x_{1,\eta} - x_1 x_{3,\eta} \\ x_{1,\eta}x_{2,\zeta} - x_{2,\eta}x_{1,\zeta} & x_2 x_{1,\zeta} - x_1 x_{2,\zeta} & x_1 x_{2,\eta} - x_2 x_{1,\eta} \end{bmatrix} \,, \tag{3.17}$$

with $x_i = x_i(\eta,\zeta)$.

To obtain a different representation of $\mathbf{J}^{-1}$, the base vectors of the transformed coordinate system $(\mathbf{x}, \mathbf{x}_{,\eta}, \mathbf{x}_{,\zeta})$ of the finite element are referred (see Fig. 3.4), and the outward normal vectors to the surfaces (η,ζ), (ζ,ξ) and (ξ,η) are calculated, using the cross products of these basis vectors:

$$\mathbf{n}^\xi = \mathbf{x}_{,\eta} \times \mathbf{x}_{,\zeta} = \begin{bmatrix} x_{2,\eta}x_{3,\zeta} - x_{3,\eta}x_{2,\zeta} \\ x_{3,\eta}x_{1,\zeta} - x_{1,\eta}x_{3,\zeta} \\ x_{1,\eta}x_{2,\zeta} - x_{2,\eta}x_{1,\zeta} \end{bmatrix} \,, \tag{3.18}$$

$$\mathbf{n}^\eta = \mathbf{x}_{,\zeta} \times \mathbf{x} = \begin{bmatrix} x_3 x_{2,\zeta} - x_2 x_{3,\zeta} \\ x_1 x_{3,\zeta} - x_3 x_{1,\zeta} \\ x_2 x_{1,\zeta} - x_1 x_{2,\zeta} \end{bmatrix} \quad \text{and} \tag{3.19}$$

$$\mathbf{n}^\zeta = \mathbf{x} \times \mathbf{x}_{,\eta} = \begin{bmatrix} x_2 x_{3,\eta} - x_3 x_{2,\eta} \\ x_3 x_{1,\eta} - x_1 x_{3,\eta} \\ x_1 x_{2,\eta} - x_2 x_{1,\eta} \end{bmatrix} \,. \tag{3.20}$$

Inserting (3.18) to (3.20) into (3.17) leads to:

$$\mathbf{J}^{-1} = \frac{1}{|\mathbf{J}|} \begin{bmatrix} n_1^\xi & n_1^\eta & n_1^\zeta \\ n_2^\xi & n_2^\eta & n_2^\zeta \\ n_3^\xi & n_3^\eta & n_3^\zeta \end{bmatrix} \,. \tag{3.21}$$

Finally, inserting (3.21) into (3.16) yields:

$$\begin{bmatrix} \frac{\partial}{\partial x_1} \\ \frac{\partial}{\partial x_2} \\ \frac{\partial}{\partial x_3} \end{bmatrix} = \frac{1}{|\mathbf{J}|} \begin{bmatrix} n_1^\xi & n_1^\eta & n_1^\zeta \\ n_2^\xi & n_2^\eta & n_2^\zeta \\ n_3^\xi & n_3^\eta & n_3^\zeta \end{bmatrix} \begin{bmatrix} \frac{\partial}{\partial \xi} \\ \frac{1}{\xi}\frac{\partial}{\partial \eta} \\ \frac{1}{\xi}\frac{\partial}{\partial \zeta} \end{bmatrix}$$

$$= \frac{\mathbf{n}^\xi}{|\mathbf{J}|}\frac{\partial}{\partial \xi} + \frac{1}{\xi}\left[\frac{\mathbf{n}^\eta}{|\mathbf{J}|}\frac{\partial}{\partial \eta} + \frac{\mathbf{n}^\zeta}{|\mathbf{J}|}\frac{\partial}{\partial \zeta} \right] \,. \tag{3.22}$$

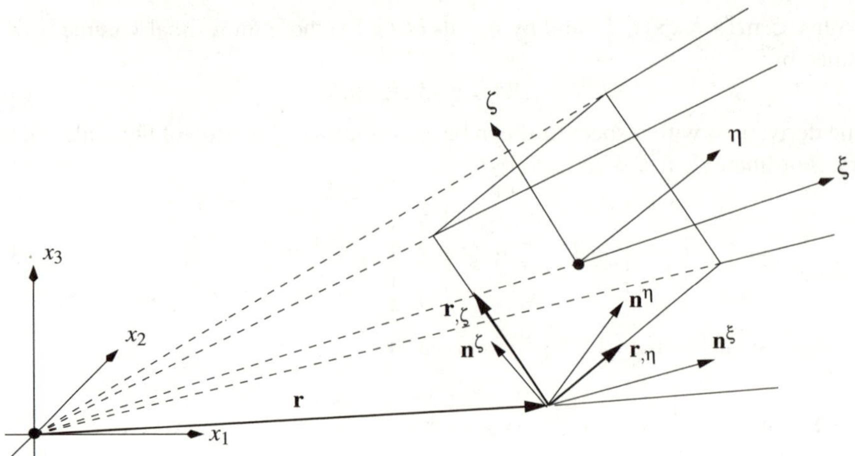

Fig. 3.4. Scaled boundary transformation for three-dimensional problems: base vectors of the transformed coordinate system and derivations of position vector

To derive the governing equations of elastodynamics with geometry in the transformed coordinate system, the scaled boundary transformation is applied to the geometry of the domain. Only the spatial coordinates are affected. The components of the displacements, strains and stresses are still defined in the original coordinate system x_i. This is analogous to the mapping of parent elements to curvilinear elements in the FEM. Thus, to transform the governing equations of elastodynamics (equations (1.1), (1.4), (3.1) and (1.7)) in the scaled boundary coordinate system, only the differential operator $\mathbf{D}$ needs to be modified.

Substituting (3.22) into (1.3) yields:

$$\mathbf{D} = \mathbf{A}_1 \frac{\partial}{\partial \xi} + \frac{1}{\xi} \left[\mathbf{A}_2 \frac{\partial}{\partial \eta} + \mathbf{A}_3 \frac{\partial}{\partial \zeta} \right] , \tag{3.23}$$

with:

$$\mathbf{A}_1 = \frac{1}{|\mathbf{J}|} \begin{bmatrix} n_1^\xi & 0 & 0 \\ 0 & n_2^\xi & 0 \\ 0 & 0 & n_3^\xi \\ 0 & n_3^\xi & n_2^\xi \\ n_3^\xi & 0 & n_1^\xi \\ n_2^\xi & n_1^\xi & 0 \end{bmatrix} , \tag{3.24}$$

$$\mathbf{A_2} = \frac{1}{|\mathbf{J}|} \begin{bmatrix} n_1^{\eta} & 0 & 0 \\ 0 & n_2^{\eta} & 0 \\ 0 & 0 & n_3^{\eta} \\ 0 & n_3^{\eta} & n_2^{\eta} \\ n_3^{\eta} & 0 & n_1^{\eta} \\ n_2^{\eta} & n_1^{\eta} & 0 \end{bmatrix} \quad \text{and} \tag{3.25}$$

$$\mathbf{A_3} = \frac{1}{|\mathbf{J}|} \begin{bmatrix} n_1^{\zeta} & 0 & 0 \\ 0 & n_2^{\zeta} & 0 \\ 0 & 0 & n_3^{\zeta} \\ 0 & n_3^{\zeta} & n_2^{\zeta} \\ n_3^{\zeta} & 0 & n_1^{\zeta} \\ n_2^{\zeta} & n_1^{\zeta} & 0 \end{bmatrix} . \tag{3.26}$$

The amplitudes of the surface tractions

$$\mathbf{t} = \begin{bmatrix} t_1 \\ t_2 \\ t_3 \end{bmatrix} \tag{3.27}$$

on any boundary with the outward unit normal vector $\mathbf{n} = n_i$ equal

$$\mathbf{t} = \begin{bmatrix} n_1 & 0 & 0 & 0 & n_3 & n_2 \\ 0 & n_2 & 0 & n_3 & 0 & n_1 \\ 0 & 0 & n_3 & n_2 & n_1 & 0 \end{bmatrix} \boldsymbol{\sigma} . \tag{3.28}$$

The traction amplitudes on the surfaces (η,ζ), (ζ,ξ) and (ξ,η), where the coordinates ξ, η and ζ, respectively, are constant, can be written as:

$$\mathbf{t}_\xi = \frac{|\mathbf{J}|}{\mathbf{n}^\xi} \mathbf{A}_1^T \boldsymbol{\sigma} \tag{3.29}$$

$$\mathbf{t}_\eta = \frac{|\mathbf{J}|}{\mathbf{n}^\eta} \mathbf{A}_2^T \boldsymbol{\sigma} \quad \text{and} \tag{3.30}$$

$$\mathbf{t}_\zeta = \frac{|\mathbf{J}|}{\mathbf{n}^\zeta} \mathbf{A}_3^T \boldsymbol{\sigma} . \tag{3.31}$$

3.3 Virtual work derivation of the scaled boundary finite element equation in displacements for the frequency domain

After transforming to scaled boundary coordinates, a virtual work formulation is applied, developed along similar lines as in the classical finite element method. It will also be demonstrated that the equivalent nodal forces are obtained directly from the virtual work statement. The differential equations of motion are formulated as far as the geometry is concerned in the scaled boundary coordinates. Following the derivation by Wolf [148], body loads are present, but for the sake of simplicity, no non-zero prescribed surface tractions on the boundary Γ_t^0 (see Fig. 3.1) are applied.

The displacement amplitudes of the finite element on the boundary ($\xi = 1$) are interpolated using shape functions $\mathbf{N}(\eta, \zeta)$, which are equal to the mapping functions (3.2). The same shape functions apply with the displacement vectors $\mathbf{u}(\xi)$ for all surfaces Γ_ξ with a constant ξ:

$$\mathbf{u}(\xi, \eta, \zeta) = \mathbf{N}(\eta, \zeta)\mathbf{u}(\xi) \,. \tag{3.32}$$

The strains $\boldsymbol{\varepsilon}$ and the stresses $\boldsymbol{\sigma}$ in dependency of the scaled boundary coordinates can be obtained, using (3.32):

$$\boldsymbol{\varepsilon}(\xi, \eta, \zeta) = \mathbf{B}_1 \mathbf{u}(\xi)_{,\xi} + \frac{1}{\xi}\mathbf{B}_2 \mathbf{u}(\xi) \ \text{ and} \tag{3.33}$$

$$\boldsymbol{\sigma}(\xi, \eta, \zeta) = \mathbf{E}\left(\mathbf{B}_1 \mathbf{u}(\xi)_{,\xi} + \frac{1}{\xi}\mathbf{B}_2 \mathbf{u}(\xi)\right) \,, \tag{3.34}$$

with

$$\mathbf{B}_1 = \mathbf{A}_1 \mathbf{N}(\eta, \zeta) \ \text{ and} \tag{3.35}$$
$$\mathbf{B}_2 = \mathbf{A}_2 \mathbf{N}(\eta, \zeta)_{,\eta} + \mathbf{A}_3 \mathbf{N}(\eta, \zeta)_{,\zeta} \,, \tag{3.36}$$

where $\mathbf{A}_1$, $\mathbf{A}_2$ and $\mathbf{A}_3$ are determined by equations (3.24) to (3.26).

The virtual displacement field is formed using the shape functions $\mathbf{N}(\eta, \zeta)$ to interpolate between nodes in the circumferential directions. The amplitudes of the virtual displacements are of the form

$$\delta\mathbf{u}(\xi, \eta, \zeta) = \mathbf{N}(\eta, \zeta)\delta\mathbf{u}(\xi) \,, \tag{3.37}$$

which is analogous to equation (3.32). The virtual state is denoted with the δ symbol. The amplitudes of the corresponding virtual strains are specified as

$$\delta\boldsymbol{\varepsilon}(\xi, \eta, \zeta) = \mathbf{B}_1 \delta\mathbf{u}(\xi)_\xi + \frac{1}{\xi}\mathbf{B}_2 \delta\mathbf{u}(\xi) \,. \tag{3.38}$$

With these equations, the virtual work statement in the frequency domain can be written as:

$$\underbrace{\int_V \delta\boldsymbol{\varepsilon}^T \boldsymbol{\sigma}\, dV}_{\text{internal work}} - \underbrace{\int_V \delta\mathbf{u}^T \rho\omega^2 \mathbf{u}\, dV}_{\text{work of internal loads}}$$

$$- \underbrace{\int_V \delta\mathbf{u}^T \mathbf{f}\, dV}_{\text{external work of body loads}} - \underbrace{\int_\Gamma \delta\mathbf{u}^T \mathbf{t}\, d\Gamma}_{\text{external work of body tractions}} = 0 \,, \tag{3.39}$$

formulated in scaled boundary coordinates:

$$\underbrace{\int_V \delta\boldsymbol{\varepsilon}^T(\xi,\eta,\zeta)\boldsymbol{\sigma}(\xi,\eta,\zeta)\,dV}_{\text{integral } I_1} - \underbrace{\int_V \delta\mathbf{u}^T(\xi,\eta,\zeta)\rho\omega^2\mathbf{u}(\xi,\eta,\zeta)\,dV}_{\text{integral } I_2}$$

$$-\underbrace{\int_V \delta\mathbf{u}^T(\xi,\eta,\zeta)\mathbf{f}(\xi,\eta,\zeta)\,dV}_{\text{integral } I_3} - \underbrace{\int_\Gamma \delta\mathbf{u}^T(\xi=1,\eta,\zeta)\mathbf{t}(\eta,\zeta)\,d\Gamma = 0}_{\text{integral } I_4}\ . \qquad (3.40)$$

Integral I_1

Integral I_1 of (3.40) can be rewritten with equation (3.37) and (3.38) as:

$$I_1 = \int_V \left(\mathbf{B}_1\delta\mathbf{u}_{,\xi} + \frac{1}{\xi}\mathbf{B}_2\delta\mathbf{u}\right)^T \mathbf{E}\left(\mathbf{B}_1\mathbf{u}_{,\xi} + \frac{1}{\xi}\mathbf{B}_2\mathbf{u}\right)\xi^2|\mathbf{J}|\,d\xi d\eta d\zeta\,, \qquad (3.41)$$

where $\mathbf{u} = \mathbf{u}(\xi)$ and $dV = \xi^2|\mathbf{J}|\,d\xi d\eta d\zeta$ (3.15) are inserted.

Integration by parts in the radial coordinate ξ for all terms $\delta\mathbf{u}_{,\xi}$ containing derivations of the virtual displacements with respect to ξ introduces integrals over the boundary Γ:

$$I_1 = \int_0^1 \delta\mathbf{u}_{,\xi}^T \int_\Gamma \mathbf{B}_1^T\mathbf{E}\mathbf{B}_1|\mathbf{J}|\,d\Gamma\,\xi^2\mathbf{u}_{,\xi}\,d\xi$$

$$+ \int_0^1 \delta\mathbf{u}_{,\xi}^T \int_\Gamma \mathbf{B}_1^T\mathbf{E}\mathbf{B}_2|\mathbf{J}|\,d\Gamma\,\xi\mathbf{u}\,d\xi$$

$$+ \int_0^1 \delta\mathbf{u}^T \int_\Gamma \mathbf{B}_2^T\mathbf{E}\mathbf{B}_1|\mathbf{J}|\,d\Gamma\,\xi\mathbf{u}_{,\xi}\,d\xi$$

$$+ \int_0^1 \delta\mathbf{u}^T \int_\Gamma \mathbf{B}_2^T\mathbf{E}\mathbf{B}_2|\mathbf{J}|\,d\Gamma\,\mathbf{u}\,d\xi\,. \qquad (3.42)$$

Equation (3.42) can be rewritten as:

$$\delta\mathbf{u}^T \int_\Gamma \mathbf{B}_1^T \mathbf{E} \mathbf{B}_1 |\mathbf{J}| \, d\Gamma \, \xi^2 \mathbf{u}_{,\xi}|_{\xi=1}$$

$$- \int_0^1 \delta\mathbf{u}^T \int_\Gamma \mathbf{B}_1^T \mathbf{E} \mathbf{B}_1 |\mathbf{J}| \, d\Gamma \, (2\xi\mathbf{u}_{,\xi} + \xi^2 \mathbf{u}_{,\xi\xi}) \, d\xi$$

$$+ \delta\mathbf{u}^T \int_\Gamma \mathbf{B}_1^T \mathbf{E} \mathbf{B}_2 |\mathbf{J}| \, d\Gamma \, \xi\mathbf{u}|_{\xi=1}$$

$$- \int_0^1 \delta\mathbf{u}^T \int_\Gamma \mathbf{B}_1^T \mathbf{E} \mathbf{B}_2 |\mathbf{J}| \, d\Gamma \, (\mathbf{u} + \xi\mathbf{u}_{,\xi} \, d\xi$$

$$+ \int_0^1 \delta\mathbf{u}^T \int_\Gamma \mathbf{B}_2^T \mathbf{E} \mathbf{B}_1 |\mathbf{J}| \, d\Gamma \, \xi\mathbf{u}_{,\xi} \, d\xi$$

$$+ \int_0^1 \delta\mathbf{u}^T \int_\Gamma \mathbf{B}_2^T \mathbf{E} \mathbf{B}_2 |\mathbf{J}| \, d\Gamma \, \mathbf{u} \, d\xi . \tag{3.43}$$

For abbreviation, the matrices $\mathbf{C}_1$, $\mathbf{C}_2$ and $\mathbf{C}_3$ are introduced:

$$\mathbf{C}_1 = \int_\Gamma \mathbf{B}_1^T \mathbf{E} \mathbf{B}_1 |\mathbf{J}| \, d\Gamma , \tag{3.44}$$

$$\mathbf{C}_2 = \int_\Gamma \mathbf{B}_2^T \mathbf{E} \mathbf{B}_1 |\mathbf{J}| \, d\Gamma \ \text{and} \tag{3.45}$$

$$\mathbf{C}_3 = \int_\Gamma \mathbf{B}_2^T \mathbf{E} \mathbf{B}_2 |\mathbf{J}| \, d\Gamma . \tag{3.46}$$

Like in the standard finite element procedure, these element matrices can be assembled for the entire boundary. With these abbreviations, and inserting $\xi = 1$, integral I_1 finally reads:

$$I_1 = \delta\mathbf{u}^T (\xi = 1) \left(\mathbf{C}_1 \mathbf{u}_{,\xi}(\xi = 1) + \mathbf{C}_2^T \mathbf{u}(\xi = 1) \right)$$

$$- \int_0^1 \delta\mathbf{u}^T \left[\mathbf{C}_1 \xi^2 \mathbf{u}_{,\xi\xi} + (2\mathbf{C}_1 + \mathbf{C}_2^T - \mathbf{C}_2) \xi\mathbf{u}_{,\xi} + (\mathbf{C}_2^T - \mathbf{C}_3) \mathbf{u} \right] d\xi . \tag{3.47}$$

Integral I_2

Inserting equation (3.37) and (3.15) into the virtual work formulation (3.40), the second integral of (3.40) can be expressed as:

$$I_2 = -\omega^2 \int_0^1 \delta\mathbf{u}^T \int_\Gamma \mathbf{N}^T \rho \mathbf{N} |\mathbf{J}| \, d\Gamma \, \xi^2 \mathbf{u} \, d\xi \,, \tag{3.48}$$

where, similar to (3.42), the integration by parts in the radial coordinate ξ for all terms containing $\delta\mathbf{u}_{,\xi}$ applies. With the coefficient matrix $\mathbf{M}$

$$\mathbf{M} = \int_\Gamma \mathbf{N}^T \rho \mathbf{N} |\mathbf{J}| \, d\Gamma \,, \tag{3.49}$$

equation (3.48) can be rewritten as

$$I_2 = -\omega^2 \int_0^1 \delta\mathbf{u}^T \mathbf{M} \xi^2 \mathbf{u} \, d\xi \,. \tag{3.50}$$

Integral I_3

The third integral I_3 (3.40) is treated in the same manner as integrals I_1 and I_2:

$$I_3 = - \int_0^1 \delta\mathbf{u}^T \xi^2 \int_{\Gamma_\xi} \mathbf{N}^T \mathbf{f} |\mathbf{J}| \, d\eta \, d\zeta \, d\xi \,, \tag{3.51}$$

where Γ_ξ denotes a surface with constant ξ.

Introducing $\mathbf{f}_b = \mathbf{f}_b(\xi)$, which corresponds to the amplitudes of the nodal forces resulting from body loads, with

$$\mathbf{f}_b = \int_{\Gamma_\xi} \mathbf{N}^T \mathbf{f} |\mathbf{J}| \, d\eta \, d\zeta \,, \tag{3.52}$$

integral I_3 becomes:

$$I_3 = - \int_0^1 \delta\mathbf{u}^T \xi^2 \mathbf{f}_b \, d\xi \,. \tag{3.53}$$

Integral I_4

Finally, the fourth integral I_4 of (3.40) can be rewritten with equation (3.37) as

$$I_4 = -\delta\mathbf{u}^T (\xi = 1) \int_\Gamma \mathbf{N}^T \mathbf{t} \, d\Gamma \,. \tag{3.54}$$

Scaled boundary finite element equation in displacements

Substituting all integrals $I_{1\ldots4}$ (equations (3.47), (3.50), (3.53) and (3.54)) in the virtual work formulation (3.40) leads to:

$$\delta\mathbf{u}^T(\xi=1)\left(\mathbf{C}_1\mathbf{u}_{,\xi}(\xi=1)+\mathbf{C}_2^T\mathbf{u}(\xi=1)-\mathbf{f}_r\right)$$

$$-\int_0^1 \delta\mathbf{u}^T\left[\mathbf{C}_1\xi^2\mathbf{u}_{,\xi\xi}\left(2\mathbf{C}_1-\mathbf{C}_2+\mathbf{C}_2^T\right)\xi\mathbf{u}_{,\xi}+\left(\mathbf{C}_2^2-\mathbf{C}_3\right)\mathbf{u}\right.$$

$$\left.+\omega^2\mathbf{M}\xi^2\mathbf{u}+\xi^2\mathbf{f}_b\right]d\xi=0\,, \tag{3.55}$$

with

$$\mathbf{f}_r=\int_\Gamma \mathbf{N}^T\mathbf{t}\,d\Gamma\,. \tag{3.56}$$

$\mathbf{f}_r$ are the amplitudes of the equivalent nodal forces due to the boundary tractions.

To satisfy (3.55) for all $\delta\mathbf{u}$ with $0\leq\xi\leq1$, both of the following conditions must be fulfilled:

$$\mathbf{f}_r=\mathbf{C}_1\mathbf{u}_{,\xi}(\xi=1)+\mathbf{C}_2^T\mathbf{u}(\xi=1)\,, \tag{3.57}$$

which is the nodal force-displacement relationship formulated at the boundary, and

$$\mathbf{C}_1\xi^2\mathbf{u}_{,\xi\xi}\left(2\mathbf{C}_1-\mathbf{C}_2+\mathbf{C}_2^T\right)\xi\mathbf{u}_{,\xi}+\left(\mathbf{C}_2^2-\mathbf{C}_3\right)\mathbf{u}+\omega^2\mathbf{M}\xi^2\mathbf{u}+\xi^2\mathbf{f}_b=\mathbf{0}\,, \tag{3.58}$$

which represents the SBFE equation in displacements, formulated in the frequency domain for elastodynamics in three dimensions. It is valid for bounded ($0\leq\xi\leq1$) and unbounded ($1\leq\xi\leq\infty$) domains.

The next step is to formulate this equation in the dynamic stiffness matrix $\mathbf{S}^\infty$ for an unbounded medium.

3.4 Virtual work derivation of scaled boundary finite element equation in dynamic stiffness for the frequency domain

To achieve the scaled boundary finite element equation in the frequency domain, it has to be formulated in the so-called dynamic stiffness (matrix) $\mathbf{S}^\infty$.

The virtual work for any surface Γ_ξ with constant ξ can be written as

$$\mathbf{w}(\xi)^T\mathbf{f}_q(\xi)=\int_{\Gamma_\xi}\mathbf{w}^T\mathbf{t}_\xi\,d\Gamma_\xi\,, \tag{3.59}$$

where the weighting function $\mathbf{w}=\mathbf{w}(\xi,\eta,\zeta)=\mathbf{w}(\xi)\mathbf{N}(\eta,\zeta)$ and $\mathbf{f}_q(\xi)$ are the amplitudes of internal forces and $\mathbf{t}_\xi$ means the resultants of surface tractions on the surface Γ_ξ (with constant ξ).

With relation (3.3) $d\Gamma_\xi$ can be calculated with

$$d\Gamma_\xi = |\mathbf{r}_{,\eta}(\xi,\eta,\zeta) \times \mathbf{r}_{,\zeta}(\xi,\eta,\zeta)| \, d\eta d\zeta$$
$$= |\xi\mathbf{r}_{,\eta}(\eta,\zeta) \times \xi\mathbf{r}_{,\zeta}(\eta,\zeta)| \, d\eta d\zeta$$
$$= \xi^2 |\mathbf{n}_\xi| \, d\eta d\zeta \, . \tag{3.60}$$

Inserting (3.60) in (3.59) for an arbitrary $\mathbf{w}(\xi)$ results in:

$$\mathbf{f}_q(\xi) = \int_{\Gamma_\xi} \mathbf{N}^T \mathbf{t}_\xi \xi^2 |\mathbf{n}_\xi| \, d\eta d\zeta \, . \tag{3.61}$$

Substituting (3.33) into (3.29) and with respect to (3.35) equation ($\mathbf{B}_1 = \mathbf{A}_1\mathbf{N}$) (3.61) can be rewritten as:

$$\mathbf{f}_q(\xi) = \int_{\Gamma_\xi} \mathbf{B}_1^T \mathbf{E} \left(\mathbf{B}_1 \mathbf{u}_{,\xi} + \frac{1}{\xi}\mathbf{B}_2\mathbf{u} \right) \xi^2 |\mathbf{J}| \, d\eta d\zeta \, . \tag{3.62}$$

Using the coefficient matrices $\mathbf{C}_1$ and $\mathbf{C}_2$ of (3.44) and (3.45), (3.62) becomes:

$$\mathbf{f}_q(\xi) = \mathbf{C}_1 \xi^2 \mathbf{u}_{,\xi} + \mathbf{C}_2^T \xi \mathbf{u} \, . \tag{3.63}$$

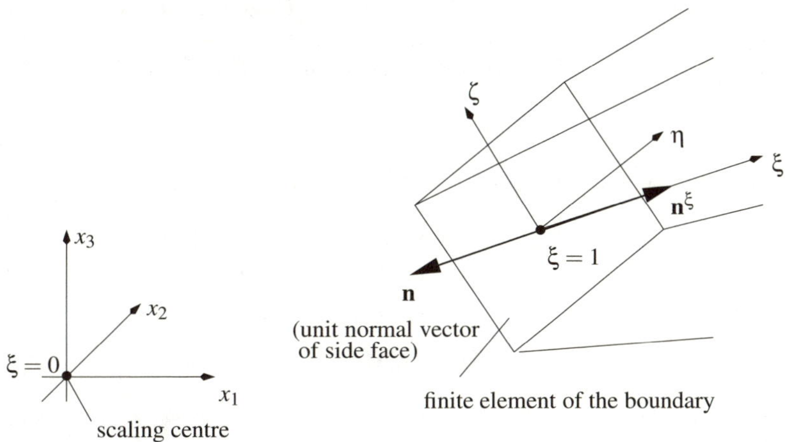

Fig. 3.5. Unit normal vector of scaled boundary finite element (unbounded case)

For an unbounded medium, $\mathbf{n}^\xi$ is opposite direction to the unit normal vector of the side-face (see Fig. 3.5), thus:

$$\mathbf{f}_r(\xi) = -\mathbf{f}_q(\xi) \, . \tag{3.64}$$

In the frequency domain, the amplitudes of the displacements $\mathbf{u}$ are related to those of the nodal force $\mathbf{f}_r(\xi)$ as

$$\mathbf{f}_r(\xi) = \mathbf{S}\mathbf{u}(\xi) - \mathbf{f}_{bs}(\xi) \, , \tag{3.65}$$

where $\mathbf{S} = \mathbf{S}(\omega,\xi)$ denotes the dynamic stiffness on a line with constant ξ and $\mathbf{f}_{bs}(\xi)$ representing the amplitudes of the nodal loads due to the body load and the surface traction. Substituting (3.63) and (3.65) into (3.64) yields:

$$-\mathbf{S}\mathbf{u} + \mathbf{f}_{bs} = \mathbf{C}_1\xi^2\mathbf{u}_{,\xi} + \mathbf{C}_2^T\xi\mathbf{u} \ . \tag{3.66}$$

Differentiating (3.66) with respect to ξ results in:

$$-\mathbf{S}_{,\xi}\mathbf{u} - \mathbf{S}\mathbf{u}_{,\xi} + \mathbf{f}_{bs,\xi} - \mathbf{C}_1\xi^2\mathbf{u}_{,\xi\xi} - (2\mathbf{C}_1 + \mathbf{C}_2^T)\xi\mathbf{u}_{,\xi} - \mathbf{C}_2^T\mathbf{u} = \mathbf{0} \ . \tag{3.67}$$

Adding the SBFE equation in displacements (3.58) and (3.67) results in:

$$-\mathbf{S}_{,\xi}\mathbf{u} + (-\mathbf{S} - \xi\mathbf{C}_2)\mathbf{u}_{,\xi} + \mathbf{f}_{bs,\xi} - \mathbf{C}_3\mathbf{u} + \omega^2\mathbf{M}\xi^2\mathbf{u} + \mathbf{q}_f = \mathbf{0} \ . \tag{3.68}$$

Solving (3.66) for $\mathbf{u}_{,\xi}$ and substituting in (3.68) multiplied by ξ results in:

$$\left[(-\mathbf{S} - \xi\mathbf{C}_2)(\xi\mathbf{C}_1)^{-1}(-\mathbf{S} - \xi\mathbf{C}_2^T) - \xi\mathbf{S}_{,\xi} - \xi\mathbf{C}_3 + \omega^2\xi^3\mathbf{M}\right]\mathbf{u}$$
$$+ \xi\mathbf{f}_{bs,\xi} + (-\mathbf{S} - \xi\mathbf{C}_2)(\xi\mathbf{C}_1)^{-1}\mathbf{f}_{bs} + \xi\mathbf{f} = \mathbf{0} \ . \tag{3.69}$$

For an arbitrary $\mathbf{u}$ the coefficient matrix $[\ldots]$ of (3.69) must vanish. This leads to

$$(-\mathbf{S} - \xi\mathbf{C}_2)(\xi\mathbf{C}_1)^{-1}(-\mathbf{S} - \xi\mathbf{C}_2^T) - \xi\mathbf{S}_{,\xi} - \xi\mathbf{C}_3 + \omega^2\xi^3\mathbf{M} = \mathbf{0} \ . \tag{3.70}$$

Introducing the dimensionless dynamic stiffness matrix $\bar{\mathbf{S}}(\omega,\xi)$, with

$$\mathbf{S}(\omega,\xi) = Gr_0\xi\bar{\mathbf{S}}(\omega,\xi) \ , \tag{3.71}$$

and substituting the non-dimensional coefficient matrices $\bar{\mathbf{C}}_1$, $\bar{\mathbf{C}}_2$, $\bar{\mathbf{C}}_3$ and $\bar{\mathbf{M}}$:

$$\mathbf{C}_1 = Gr_0\bar{\mathbf{C}}_1 \ , \tag{3.72}$$
$$\mathbf{C}_2 = Gr_0\bar{\mathbf{C}}_2 \ , \tag{3.73}$$
$$\mathbf{C}_3 = Gr_0\bar{\mathbf{C}}_3 \text{ and} \tag{3.74}$$
$$\mathbf{M} = \rho r_0^3\bar{\mathbf{M}} \tag{3.75}$$

into (3.70) yields:

$$-\bar{\mathbf{S}} - \bar{\mathbf{C}}_2\bar{\mathbf{C}}_1^{-1}(-\bar{\mathbf{S}} - \bar{\mathbf{C}}_2^T) - \bar{\mathbf{C}}_3^T - \bar{\mathbf{S}} - \xi\bar{\mathbf{S}}_{,\xi} + \left(\frac{\omega r_o}{c_s}\xi\right)^2\bar{\mathbf{M}} = \mathbf{0} \ , \tag{3.76}$$

with

$$c_s = \sqrt{\frac{G}{\rho}} \tag{3.77}$$

is the shear-wave velocity .

The coefficient of the last term of (3.76) defines the dimensionless frequency for any ξ:

$$a(\omega,\xi) = \frac{\omega r_o}{c_s}\xi \ , \tag{3.78}$$

thus $\xi\bar{\mathbf{S}}_{,\xi}$ of (3.76) can be written as

$$\xi\bar{\mathbf{S}}_{,\xi} = a\bar{S}_{,a} \; . \tag{3.79}$$

After substituting (3.78) and (3.79) into (3.77), the only independent variable in the equation is a. Hence, the dimensionless dynamic-stiffness matrix is a function of a only:

$$\bar{\mathbf{S}}(\omega,\xi) = \bar{\mathbf{S}}(a) \; . \tag{3.80}$$

The term with the derivative $a\bar{\mathbf{S}}(a)_{,a}$ can be interpreted either for varying ξ with fixed ω, or for varying ω with fixed ξ:

$$a\bar{\mathbf{S}}_{,a} = \xi\bar{\mathbf{S}}_{,\xi} = \omega\bar{\mathbf{S}}(a)_{,\omega} \; . \tag{3.81}$$

Differentiating (3.71) with respect to ξ and using (3.80) and (3.81) yields:

$$\xi\mathbf{S}_{,\xi} = \mathbf{S} + \omega\mathbf{S}_{,\omega} \; , \tag{3.82}$$

so the spatial derivative can be replaced by the frequency derivative. Substituting (3.82) into (3.69) results in

$$(-\mathbf{S} - \xi\mathbf{C}_2)(\xi\mathbf{C}_1)^{-1}(-\mathbf{S} - \xi\mathbf{C}_2^T) - \xi\mathbf{C}_3 - \mathbf{S} - \omega\mathbf{S}_{,\omega} + \omega^2\xi^3\mathbf{M} = \mathbf{0} \; . \tag{3.83}$$

For the boundary $(\xi = 1)$ the dynamic-stiffness matrix for an unbounded medium $S^{\infty}(\omega)$ is expressed as:

$$(\mathbf{S}^{\infty}(\omega) + \mathbf{C}_2)\mathbf{C}_1^{-1}(\mathbf{S}^{\infty}(\omega) + \mathbf{C}_2^T) - \mathbf{S}^{\infty}(\omega) - \omega\mathbf{S}^{\infty}(\omega)_{,\omega} - \mathbf{C}_3 + \omega^2\mathbf{M} = \mathbf{0} \; . \tag{3.84}$$

Equation (3.84) represents the SBFE equation in dynamic-stiffness formulated in the frequency domain for elastodynamics in three dimensions. It is a non-linear first-order ordinary differential equation, with the frequency ω as the independent variable, and $\mathbf{S}^{\infty}(\omega)$ stands for the unknown dynamic-stiffness matrix for an unbounded medium.

3.5 Derivation of the acceleration unit-impulse response matrix (time domain)

Now, for obtaining the acceleration unit impulse matrix $\mathbf{M}^{\infty}$ which is mandatory for a time domain analysis, (3.84) has to be transferred into the time domain.

Scaled boundary finite element equation in the time domain

The interaction force-acceleration relationship can be formulated as:

$$\mathbf{f}_r(t) = \int_0^t \mathbf{M}^{\infty}(t - \tau)\ddot{\mathbf{u}}(\tau)\,d\tau \; , \tag{3.85}$$

where $\mathbf{M}^{\infty}(t)$ is the acceleration unit-impulse response matrix in the time domain. Transferring (3.85) to the frequency domain and with

$$\ddot{\mathbf{u}}(\omega) = (i\omega)^2 \mathbf{u}(\omega) \tag{3.86}$$

results in:

$$\mathbf{f}_r(\omega) = \mathbf{M}^{\infty}(\omega)(i\omega)^2 \mathbf{u}(\omega) \;. \tag{3.87}$$

Here, $\mathbf{M}^{\infty}(\omega)$ is the acceleration unit-impulse response matrix in the frequency domain. $\mathbf{M}^{\infty}(t)$ and $\mathbf{M}^{\infty}(\omega)$ form a *Fourier* transform pair. With the interaction force-displacement relationship in the frequency domain is given by

$$\mathbf{f}_r(\omega) = \mathbf{S}^{\infty}(\omega)\mathbf{u}(\omega) \;, \tag{3.88}$$

where $\mathbf{u}(\omega)$ denotes the displacement amplitudes, and $\mathbf{f}_r(\omega)$ the interaction force amplitudes. With (3.88) a relationship between the acceleration and the displacement unit-impulse response matrix can be stated:

$$\mathbf{M}^{\infty}(\omega) = \frac{\mathbf{S}^{\infty}(\omega)}{(i\omega)^2} \;. \tag{3.89}$$

With (3.89) the displacement stiffness matrix $\mathbf{S}^{\infty}(\omega)$ of (3.84) can be transformed into the acceleration dynamic stiffness matrix $\mathbf{M}^{\infty}(\omega)$ by dividing this equation by $(i\omega)^4$:

$$\mathbf{M}^{\infty}(\omega)\mathbf{C}_1^{-1}\mathbf{M}^{\infty}(\omega) + \mathbf{C}_2\mathbf{C}_1^{-1}\frac{\mathbf{M}^{\infty}(\omega)}{(i\omega)^2} + \frac{\mathbf{M}^{\infty}(\omega)}{(i\omega)^2}\mathbf{C}_1^{-1}\mathbf{C}_2^T - \frac{\mathbf{M}^{\infty}(\omega)}{(i\omega)^2}$$
$$+ \frac{1}{\omega}\mathbf{M}_{,\omega}^{\infty}(\omega) - \frac{1}{(i\omega)^4}\left[\mathbf{C}_3 - \mathbf{C}_2\mathbf{C}_1^{-1}\mathbf{C}_2^T\right] - \frac{1}{(i\omega)^2}\mathbf{M} = \mathbf{0} \;. \tag{3.90}$$

Applying the inverse Fourier transformation, (3.90) results in:

$$\int_0^t \mathbf{M}^{\infty}(t-\tau)\mathbf{C}_1^{-1}\mathbf{M}^{\infty}(\tau)\,d\tau$$
$$+ (\mathbf{C}_2\mathbf{C}_1^{-1} - 2)\int_0^t\int_0^\tau \mathbf{M}^{\infty}(\tau')\,d\tau'd\tau + \int_0^t\int_0^\tau \mathbf{M}^{\infty}(\tau')\,d\tau'd\tau(\mathbf{C}_1^{-1}\mathbf{C}_2^T - 2)$$
$$+ t\int_0^t \mathbf{M}^{\infty}(\tau)\,d\tau - \frac{t^3}{6}\left[\mathbf{C}_3 - \mathbf{C}_2\mathbf{C}_1^{-1}\mathbf{C}_1^T\ \ H(t) - t\mathbf{M}H(t) = \mathbf{0} \;, \tag{3.91}$$

where

$$\mathcal{F}^{-1}\left[\frac{1}{\omega}\mathbf{M}^{\infty}(\omega)_{,\omega}\right] = \int_0^t \tau\mathbf{M}^{\infty}(\tau)\,d\tau$$
$$= t\int_0^t \mathbf{M}^{\infty}(\tau)d\tau - \int_0^t\int_0^\tau \mathbf{M}^{\infty}(\tau')\,d\tau'd\tau \;. \tag{3.92}$$

$\mathcal{F}^{-1}[\ldots]$ denotes the inverse Fourier transform of $[\ldots]$.

With the unknown matrix

$$\tilde{\mathbf{M}}^{\infty}(t) = \mathbf{U}^{-1\,T}\mathbf{M}^{\infty}(t)\mathbf{U}^{-1}\,, \tag{3.93}$$

and the coefficient matrices

$$\tilde{\mathbf{C}}_1 = \mathbf{U}^T\mathbf{U}\,, \tag{3.94}$$

$$\tilde{\mathbf{C}}_2 = \mathbf{U}^{-1\,T}\mathbf{C}_2\mathbf{U}^{-1} - 2\mathbf{I}\,, \tag{3.95}$$

$$\tilde{\mathbf{C}}_3 = \mathbf{U}^{-1\,T}\left(\mathbf{C}_3 - \mathbf{C}_2\mathbf{C}_1^{-1}\tilde{\mathbf{C}}_2^T\right)\mathbf{U}^{-1}\quad\text{and} \tag{3.96}$$

$$\tilde{\mathbf{M}} = \mathbf{U}^{-1\,T}\mathbf{M}\mathbf{U}^{-1}\,, \tag{3.97}$$

where $\mathbf{U}$ results from a Cholesky decomposition of $\tilde{\mathbf{C}}_1$ and $H(t)$ means the Heaviside step function , the SBFE equation in the time domain may be obtained from (3.92):

$$\underbrace{\int_0^t \tilde{\mathbf{M}}^{\infty}(t-\tau)\tilde{\mathbf{M}}^{\infty}(\tau)\,d\tau}_{\text{integral }J_1} + \tilde{\mathbf{C}}_2\underbrace{\int_0^t\int_0^{\tau}\tilde{\mathbf{M}}^{\infty}(\tau')\,d\tau'd\tau}_{\text{integral }J_2}$$

$$+ \underbrace{\int_0^t\int_0^{\tau}\tilde{\mathbf{M}}^{\infty}(\tau')\,d\tau'd\tau\;\tilde{\mathbf{C}}_2^T}_{\text{integral }J_2} + \underbrace{t\int_0^t\tilde{\mathbf{M}}^{\infty}(\tau)\,d\tau}_{\text{integral }J_3} - \frac{t^3}{6}\tilde{\mathbf{C}}_3H(t) - t\tilde{\mathbf{M}}H(t) = \mathbf{0}\,. \tag{3.98}$$

After a discretisation with respect to time, one can determine an approximation of $\tilde{\mathbf{M}}^{\infty}(t)$ from (3.98). Finally, the desired acceleration unit-impulse response matrix $\mathbf{M}^{\infty}(t)$ is gained by

$$\mathbf{M}^{\infty}(t) = \mathbf{U}^T\tilde{\mathbf{M}}^{\infty}(t)\mathbf{U}\,. \tag{3.99}$$

3.5.1 Time discretisation

After discretisation with respect to time, (3.98) yields an equation for the acceleration unit-impulse response matrix at each time-station i. It is assumed that the response matrix is piecewise constant over each time-step. When calculating the interaction force using (3.85), this assumption is the same as for the acceleration in the constant acceleration Newmark method .

The integral J_1 of (3.98):

$$J_1 = \int_0^t \tilde{\mathbf{M}}^{\infty}(t-\tau)\tilde{\mathbf{M}}^{\infty}(\tau)\,d\tau \tag{3.100}$$

is discretised as:

$$J_1(t_i) = \Delta t \sum_{j=1}^{i} \tilde{\mathbf{M}}^{\infty}_{i-j+1}\tilde{\mathbf{M}}^{\infty}_{j}\,, \tag{3.101}$$

while the integral J_2

$$J_2 = \int\limits_0^t \int\limits_0^\tau \tilde{\mathbf{M}}^\infty(\tau')\, d\tau'\, d\tau \tag{3.102}$$

is discretised as

$$J_2(t_i) = \int\limits_0^{i\Delta t} \int\limits_0^\tau \tilde{\mathbf{M}}^\infty(\tau')\, d\tau'\, d\tau = J_2(t_{i-1}) + \Delta t J_3(t_{i-1}) + \frac{\Delta t^2}{2}\tilde{\mathbf{M}}_i^\infty \tag{3.103}$$

and, finally, the integral J_3

$$J_3 = \int\limits_0^t \tilde{\mathbf{M}}^\infty(\tau)\, d\tau \tag{3.104}$$

is likewise discretised as:

$$J_3(t_i) = \int\limits_0^{i\Delta t} \tilde{\mathbf{M}}^\infty(\tau)\, d\tau = J_1(t_{i-1}) + \Delta t \tilde{\mathbf{M}}_i^\infty\ . \tag{3.105}$$

First time-step

The convolution integral of (3.98) leads to a quadratic equation in the unknown matrix $\tilde{\mathbf{M}}^\infty$ for the first time step t_1 and a special treatment is necessary. For the first time-step, the integrals J_1, J_2 and J_3 in discrete form are:

$$J_1(t_1) = \Delta t \tilde{\mathbf{M}}_1^\infty\ , \tag{3.106}$$

$$J_2(t_1) = \frac{\Delta t^2}{2}\tilde{\mathbf{M}}_1^\infty\ \text{ and} \tag{3.107}$$

$$J_3(t_1) = \Delta t \tilde{\mathbf{M}}_1^\infty \tilde{\mathbf{M}}_1^\infty = \Delta t (\tilde{\mathbf{M}}_1^\infty)^2\ . \tag{3.108}$$

Inserting J_1, J_2 and J_3 for the first time-step in (3.98) and dividing by Δt results in:

$$(\tilde{\mathbf{M}}_1^\infty)^2 + \tilde{\mathbf{C}}_2 \frac{\Delta t^2}{2}\tilde{\mathbf{M}}_1^\infty + \frac{\Delta t^2}{2}\tilde{\mathbf{M}}_1^\infty \tilde{\mathbf{C}}_2^T + t\tilde{\mathbf{M}}_1^\infty - \frac{\Delta t^2}{6}\tilde{\mathbf{C}}_3 - \tilde{\mathbf{M}} = \mathbf{0}\ . \tag{3.109}$$

Rearranging (3.109) results in the algebraic *Riccati equation* in $\tilde{\mathbf{M}}_1^\infty$, where a solution procedure exists:

$$(\tilde{\mathbf{M}}_1^\infty)^2 + \frac{\Delta t}{2}\left(\tilde{\mathbf{C}}_2 + \mathbf{I}\right)\tilde{\mathbf{M}}_1^\infty + \tilde{\mathbf{M}}_1^\infty \frac{\Delta t}{2}\left(\tilde{\mathbf{C}}_2 + \mathbf{I}\right) - \frac{\Delta t^2}{6}\tilde{\mathbf{C}}_3 - \tilde{\mathbf{M}} = \mathbf{0}\ . \tag{3.110}$$

An efficient solution procedure applies the Schur factorisation . Equation (3.110) is solved introducing the matrix

$$\mathbf{Z} = \begin{bmatrix} -\frac{\Delta t}{2}\left(\tilde{\mathbf{C}}_2^T + \mathbf{I}\right) & -\mathbf{I} \\ -\frac{\Delta t^2}{6}\tilde{\mathbf{C}}_2 - \mathbf{M} & \frac{\Delta t}{2}\left(\tilde{\mathbf{C}}_2 + \mathbf{I}\right) \end{bmatrix}\ . \tag{3.111}$$

A real orthogonal transformation $\mathbf{V}$ is applied to $\mathbf{Z}$ which yields the real Schur form with the matrix $\mathbf{S}$ in the quasi-upper triangular form:

$$\mathbf{V}^T \mathbf{Z} \mathbf{V} = \mathbf{S} = \begin{bmatrix} \mathbf{S}_{11} & \mathbf{S}_{12} \\ 0 & \mathbf{S}_{22} \end{bmatrix} . \tag{3.112}$$

$\mathbf{S}$ is arranged in such manner that the real parts of the eigenvalues of $\mathbf{S}_{11}$ are negative and those of $\mathbf{S}_{22}$ are positive. $\mathbf{V}$ is partitioned conformably as

$$\mathbf{V} = \begin{bmatrix} \mathbf{V}_{11} & \mathbf{V}_{12} \\ \mathbf{V}_{21} & \mathbf{V}_{22} \end{bmatrix} . \tag{3.113}$$

The solution of (3.110) is formulated as

$$\tilde{\mathbf{M}}_1^\infty = \mathbf{V}_{21} \mathbf{V}_{11}^{-1} . \tag{3.114}$$

The Schur factorisation of (3.112) is widely applied in solving eigenvalue problems. Efficient and accurate programmes in the public domain exist, i.e. LAPACK [122].

The acceleration unit-impulse response matrix at the first time-station follows from (3.99) as

$$\mathbf{M}_1^\infty = \mathbf{U}^T \tilde{\mathbf{M}}_1^\infty \mathbf{U} . \tag{3.115}$$

i-th time-step

For $i \geq 2$ the convolution integral term in (3.98) is linear in the unknown $\tilde{\mathbf{M}}_i^\infty$. Discretising (3.98) leads to the equation for the transformed acceleration unit-impulse matrix $\tilde{\mathbf{M}}_i^\infty$. Inserting the discretised integrals $J_1(t_i)$, $J_2(t_i)$ and $J_3(t_i)$ in (3.98) and a rearrangement leads to:

$$\left[\tilde{\mathbf{M}}_1^\infty + \frac{\Delta t}{2} \tilde{\mathbf{C}}_2 \right] \tilde{\mathbf{M}}_i^\infty + \tilde{\mathbf{M}}_i^\infty \left[\tilde{\mathbf{M}}_1^\infty + \frac{\Delta t}{2} \tilde{\mathbf{C}}_2^T \right] + t \tilde{\mathbf{M}}_i^\infty =$$
$$- \frac{J_1(t_{i-1})}{\Delta t} - \tilde{\mathbf{C}}_2 \left[\frac{J_2(t_{i-1})}{\Delta t} + J_3(t_{i-1}) \right] - \left[\frac{J_2(t_{i-1})}{\Delta t} + J_3(t_{i-1}) \right] \tilde{\mathbf{C}}_2^T$$
$$+ \frac{t^3}{6\Delta t} \tilde{\mathbf{C}}_3 + \frac{t}{\Delta t} \left[\tilde{\mathbf{M}} - J_3(t_{i-1}) \right] . \tag{3.116}$$

(3.116) is the *Lyapunov equation* in the form

$$\mathbf{A}\mathbf{X} + \mathbf{X}\mathbf{A}^T + t\mathbf{X} = \mathbf{C} \tag{3.117}$$

with the unknown $\mathbf{X} = \tilde{\mathbf{M}}_i^\infty$. A summary of the solution procedure for

$$\mathbf{A}\mathbf{X} + \mathbf{X}\mathbf{A}^T = \mathbf{C} \tag{3.118}$$

described in [12] follows. A Schur factorisation is applied to $\mathbf{A}$:

$$\mathbf{V}^T \mathbf{A} \mathbf{V} = \mathbf{S} \tag{3.119}$$

with the orthogonal transformation matrix $\mathbf{V}$, which results in the quasi-upper triangular matrix $\mathbf{S}$. (3.118) is pre-multiplied by $\mathbf{V}^T$ and post-multiplied by $\mathbf{V}$ yielding

$$\mathbf{S}\mathbf{Y} + \mathbf{Y}\mathbf{S}^T + t\mathbf{Y} = \mathbf{Y}^T\mathbf{C}\mathbf{V} \,, \tag{3.120}$$

where $\mathbf{Y} = \mathbf{V}^T\mathbf{X}\mathbf{V}$. The orthogonal matrix $\mathbf{V}$ satisfies $\mathbf{V}\mathbf{V}^T = \mathbf{V}^T\mathbf{V} = \mathbf{I}$. (3.120) is rewritten as

$$\left[\mathbf{S} + \frac{1}{2}t\mathbf{I}\right]\mathbf{Y} + \mathbf{Y}\left[\mathbf{S} + \frac{1}{2}t\mathbf{I}\right] = \mathbf{V}^T\mathbf{C}\mathbf{V} \,. \tag{3.121}$$

The matrix $\mathbf{S} + \frac{1}{2}t\mathbf{I}$ presents the real Schur form. This permits (3.121) to be solved for each time-station t_i successively leading to $\mathbf{Y}$, while $\mathbf{X}$ can be computed with

$$\mathbf{X} = \mathbf{V}\mathbf{Y}\mathbf{V}^T \,. \tag{3.122}$$

Matrix $\mathbf{A}$ of (3.117) to (3.119) is independent of the time-stations, thus the Schur factorisation has to be performed only once. The term $t\mathbf{X}$ does not increase the computational effort.

Finally, the acceleration unit-impulse response matrix is calculated by:

$$\mathbf{M}_i^\infty = \mathbf{U}^T\tilde{\mathbf{M}}_i^\infty\mathbf{U} \,. \tag{3.123}$$

$\mathbf{M}_i^\infty$ is, in general, fully populated, but symmetric.

3.6 Numerical implementation (near-field/far-field coupling)

For elastodynamic problems in the time domain a short overview of the numerical implementation is given.

For a numerical treatment in the time domain, a time discretisation of the convolution integral (3.85) is needed. If a piecewise constant approximation of the acceleration unit-impulse response matrix is assumed, i.e.,

$$\mathbf{M}^\infty(t) = \begin{cases} \mathbf{M}_0^\infty & t \in [0;\Delta t] \\ \mathbf{M}_1^\infty & t \in [\Delta t; 2\Delta t] \\ \;\;\vdots \\ \mathbf{M}_{n-1}^\infty & t \in [(n-1)\Delta t; n\Delta t] \end{cases} \tag{3.124}$$

the discrete form of (3.85) can be written as:

$$\mathbf{f}_r(t_n) = \sum_{j=1}^{n}\mathbf{M}_{n-j}^\infty\int_{(j-1)\Delta t}^{j\Delta t}\ddot{\mathbf{u}}(\tau)\,d\tau = \sum_{j=1}^{n}\mathbf{M}_{n-j}^\infty(\dot{\mathbf{u}}_j - \dot{\mathbf{u}}_{j-1}) \,. \tag{3.125}$$

Introducing the γ-parameter of the Hilber-Hughes-Taylor implicit time integration scheme (see Sec. 1.4.2), and separating the unknown acceleration vector $\ddot{\mathbf{u}}_n$, the calculation of $\mathbf{f}_r(t_i)$ can be performed with the following equation:

$$\mathbf{f}_r(t_n) = \gamma \Delta t \mathbf{M}_0^\infty \ddot{\mathbf{u}}_n + \sum_{j=1}^{n-1} \mathbf{M}_{n-j}^\infty (\dot{\mathbf{u}}_j - \dot{\mathbf{u}}_{j-1}) \ . \qquad (3.126)$$

For large i (i.e., a long simulation time) the direct solution of (3.126) is very time consuming, but an approximation in time (presented in the next subsection) leads to a fast recursive algorithm.

The standard finite element equation for elastodynamics in the time domain

$$\mathbf{Ku}(t) + \mathbf{C}\dot{\mathbf{u}}(t) + \mathbf{M}\ddot{\mathbf{u}}(t) = \mathbf{f}(t) \ , \qquad (3.127)$$

with $\mathbf{K}$, $\mathbf{C}$ and $\mathbf{M}$ being the stiffness, damping and mass matrix, respectively, can be split into a structure (near-field) and boundary part:

$$\begin{bmatrix} \mathbf{K}_{ss} & \mathbf{K}_{sb} \\ \mathbf{K}_{bs} & \mathbf{K}_{bb} \end{bmatrix} \begin{bmatrix} \mathbf{u}_s(t) \\ \mathbf{u}_b(t) \end{bmatrix} + \begin{bmatrix} \mathbf{C}_{ss} & \mathbf{C}_{sb} \\ \mathbf{C}_{bs} & \mathbf{C}_{bb} \end{bmatrix} \begin{bmatrix} \dot{\mathbf{u}}_s(t) \\ \dot{\mathbf{u}}_b(t) \end{bmatrix}$$
$$+ \begin{bmatrix} \mathbf{M}_{ss} & \mathbf{M}_{sb} \\ \mathbf{M}_{bs} & \mathbf{M}_{bb} + \gamma \Delta t \mathbf{M}_0^\infty \end{bmatrix} \begin{bmatrix} \ddot{\mathbf{u}}_s(t) \\ \ddot{\mathbf{u}}_b(t) \end{bmatrix} = \begin{bmatrix} \mathbf{f}_s(t) \\ \mathbf{f}_b(t) - \mathbf{f}_r(t) \end{bmatrix} , \qquad (3.128)$$

where the subscript s stands for structure (which is the finite element region, which may also include a portion of the ground for soil-structure interaction problems) and b stands for the boundary, which is congruent with the SBFE discretisation. $\mathbf{f}_r(t)$ are the time-dependent forces on the near-field/far-field interface due to the influence of the infinite space. These forces are calculated via the convolution integral (3.85).

3.7 Improvement of efficiency

A drawback of the SBFEM is the computational time and storage consuming non-locality in time and space. The time non-locality results in a convolution integral (3.85), which has to be solved. Here, for the n-th time-step $n-1$ matrix vector multiplications with the fully populated unit-impulse acceleration influence matrices have to be performed. Therefore, the computational effort grows quadratic in dependence of the simulation time. To reduce this computation effort to a linear time dependency, a recursive algorithm is introduced in the following.

Furthermore, an immense storage consummation when executing the SBFEM results from fully populated unit-impulse acceleration influence matrices. While for each time-station a different influence matrix has to be recorded, the requirements of storage are unacceptable for large near-field/far-field interfaces (which results in large influence matrices) or for long simulation times. Here, a storage reduction method is presented, which leads to a banded structure of the influence matrices, like the stiffness matrices in finite elements.

The combined use of recursive algorithm and storage reduction method improves the efficiency of the SBFEM significantly.

3.7.1 Convolution (reduction of non-locality in time)

The calculation of the interaction force vector on the near-field/far-field interface $\mathbf{f}_r(t)$ for long simulation periods is computing time consuming, thus a recursive algorithm is elaborated which leads to a reduction of computational effort.

A closer look at the acceleration unit-impulse matrix $\mathbf{M}^\infty$ shows that the entries of matrices $\mathbf{M}^\infty(t_1)$, $\mathbf{M}^\infty(t_2)$, … grow linear from a certain time-step t_m (see Fig. 3.6).

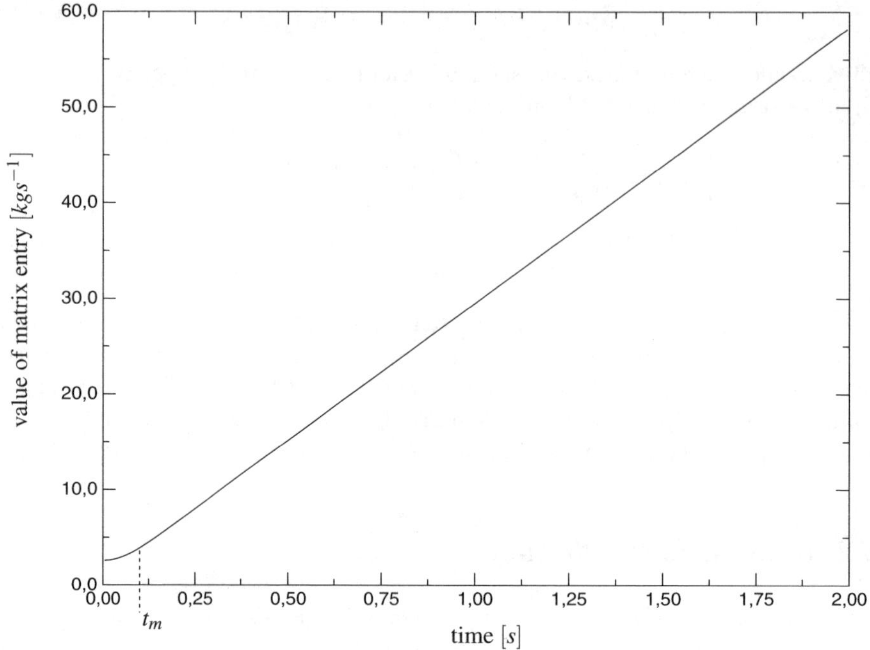

Fig. 3.6. Growth of a certain entry located on the main diagonal of the unit-impulse influence matrix in dependency on the simulation time

Hence, the acceleration unit-impulse matrix can be decomposed (from time-step t_m on) as

$$\mathbf{M}^\infty(t_i) = \mathbf{T}^\infty t_i + \mathbf{C}^\infty \tag{3.129}$$

with the matrix $\mathbf{T}^\infty$ of gradients $\left(\frac{\Delta \mathbf{M}^\infty}{\Delta t}\right)$ and the constant matrix $\mathbf{C}^\infty$. For equidistant time-steps $t_i = i\Delta t$, the calculation of $\mathbf{T}^\infty$ is reduced to

$$\mathbf{T}^\infty = \left(\mathbf{M}^\infty_{m+1} - \mathbf{M}^\infty_m\right) \quad \text{with} \quad \mathbf{M}^\infty_m = \mathbf{M}^\infty(t_m). \tag{3.130}$$

If a linear behaviour of matrix entries of $\mathbf{M}^\infty$ is assumed and $\tilde{\mathbf{f}}_r(t_n)$ is introduced, following (3.126):

$$\mathbf{f}_r(t_n) = \gamma \Delta t \mathbf{M}^\infty_0 \ddot{\mathbf{u}}_n + \sum_{j=1}^{n-1} \mathbf{M}^\infty_{n-j} \left(\dot{\mathbf{u}}_j - \dot{\mathbf{u}}_{j-1}\right) = \gamma \Delta t \mathbf{M}^\infty_0 \ddot{\mathbf{u}}_n + \tilde{\mathbf{f}}_r(t_n), \tag{3.131}$$

$\tilde{\mathbf{f}}_r(t_n)$ can be split into a non-linear term (for time-steps $t_i, 0 \le i \le m$) and a linear term (time-steps $t_i,\ i > m$):

$$\tilde{\mathbf{f}}_r(t_n) = \tilde{\mathbf{f}}_r(t_n)^{lin} + \tilde{\mathbf{f}}_r(t_n)^{nonlin}$$

$$= \sum_{j=1}^{n+1-m} \mathbf{M}^{\infty}_{n-j+1}(\dot{\mathbf{u}}_j - \dot{\mathbf{u}}_{j-1}) + \sum_{j=n+2-m}^{n-1} \mathbf{M}^{\infty}_{n-j+1}(\dot{\mathbf{u}}_j - \dot{\mathbf{u}}_{j-1}) . \quad (3.132)$$

With (3.129), the linear part of (3.132) can be written as

$$\tilde{\mathbf{f}}_r(t_n)^{lin} = \sum_{j=1}^{n+1-m} \left[\mathbf{T}^{\infty}\,(n-j+1) + \mathbf{C}^{\infty} \right] (\dot{\mathbf{u}}_j - \dot{\mathbf{u}}_{j-1})$$

$$= \left[\mathbf{T}^{\infty}\,m + \mathbf{C}^{\infty} \right] (\dot{\mathbf{u}}_{n+1-m} - \dot{\mathbf{u}}_{n-m})$$

$$+ \sum_{j=1}^{n-m} \left[\mathbf{T}^{\infty}\,(n-j+1) + \mathbf{C}^{\infty} \right] (\dot{\mathbf{u}}_j - \dot{\mathbf{u}}_{j-1})$$

$$= \mathbf{M}^{\infty}_m (\dot{\mathbf{u}}_{n+1-m} - \dot{\mathbf{u}}_{n-m})$$

$$+ \sum_{j=1}^{n-m} \left[\mathbf{T}^{\infty}\,(n-j+1) + \mathbf{C}^{\infty} \right] (\dot{\mathbf{u}}_j - \dot{\mathbf{u}}_{j-1}) . \quad (3.133)$$

For a recursive algorithm, $\tilde{\mathbf{f}}_r(t_{n-1})^{lin}$ is needed:

$$\tilde{\mathbf{f}}_r(t_{n-1})^{lin} = \sum_{j=1}^{n-m} \left[\mathbf{T}^{\infty}\,(n-j) + \mathbf{C}^{\infty} \right] (\dot{\mathbf{u}}_j - \dot{\mathbf{u}}_{j-1}). \quad (3.134)$$

With this equation, the difference $\tilde{\mathbf{f}}_r(t_n)^{lin} - \tilde{\mathbf{r}}^{lin}_{n-1}$ can be rewritten, and the recursive formulation reads finally:

$$\tilde{\mathbf{f}}_r(t_n)^{lin} = \tilde{\mathbf{f}}_r(t_{n-1})^{lin} + \mathbf{M}^{\infty}_m (\dot{\mathbf{u}}_{n+1-m} - \dot{\mathbf{u}}_{n-m}) + \mathbf{T}^{\infty} (\dot{\mathbf{u}}_{n-m} - \dot{\mathbf{u}}_0) . \quad (3.135)$$

The required relevant number of operations (multiplications) for n time-steps is reduced from

$$\mathrm{op}(n) = \frac{3}{2} DOF^2 n^2 , \quad (3.136)$$

(where DOF means degrees of freedom of the near-field/far-field interface for the direct calculation) to

$$\mathrm{op}(m) = O(\psi)\mathrm{op}(n) = 3 DOF^2 nm - \frac{3}{2} DOF^2 m^2 , \quad \text{with } \psi = \frac{m}{n} . \quad (3.137)$$

The relative computational effort $O(\psi)$ in dependence on $\psi = \frac{m}{n}$ is shown in Fig. 3.7, e.g., when $n = 5000$ time-steps should be calculated, and $m = 500$ is evaluated, it results in $\psi = \frac{m}{n} = 0.1$, and $O(\psi) = 0.19$ (see (3.137) or Fig. 3.7). Hence, the reduction of necessary operations in this case is 81%.

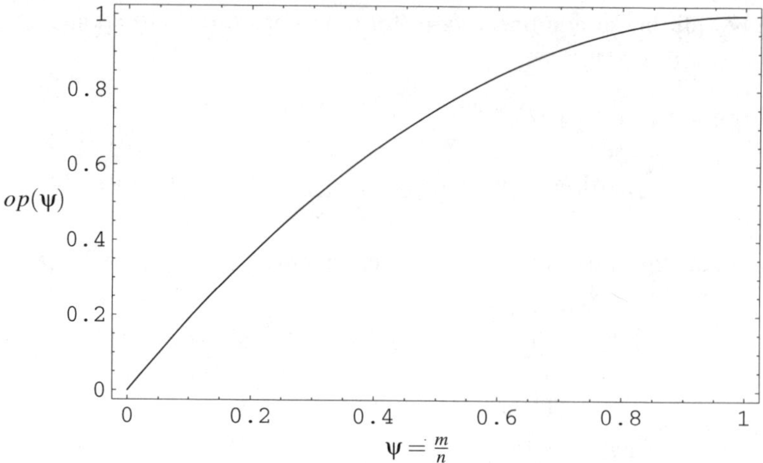

Fig. 3.7. Computational effort in dependence on ψ

3.7.2 Influence Matrices (reduction of non-locality in space)

Another way to improve the efficiency of the SBFEM is to reduce the non-locality in space. The influence matrices $\mathbf{M}^\infty(t)$ of the SBFE equation in the time domain (3.98) are symmetric, but fully populated, i.e., each node on the near-field/far-field interface has an influence on every other node on the interface. This is clearly true for small, compact interfaces, but for large or long interfaces, a significant error will not be introduced to the calculation, if neglecting the influence between remote nodes.

To neglect the influence distance nodes have on each other, a zero-element threshold is introduced, where a matrix element is set to zero if it is less equal this threshold:

$$\text{if } m_{ij} \leq \varepsilon_z \rightarrow m_{ij} := 0 \tag{3.138}$$

with m_{ij} is a matrix entry of ith row and jth column of a influence matrix $\mathbf{M}_n^\infty$. Figs. 3.8 to 3.10 show influence matrices of the near-field/far-field interface for different zero-element thresholds, where matrix entries $m_{ij} \neq 0$ are marked with a black dot.

To achieve a banded structure of the matrix (like in Figs. 3.8 to 3.10), the entries of the influence matrices have to be reordered, e.g., with the Cuthill/McKee algorithm [41]. After rearranging and introducing a zero-element threshold, the matrix population is significantly reduced and shows a banded structure. Thus effective storage and calculation techniques, like sparse matrix vector multiplication algorithms, are applicable. Consequently, introducing this zero-element threshold reduces the storage and CPU time consumption significantly, without loosing accuracy. For further details, see Part II (Applications), or [91, 90].

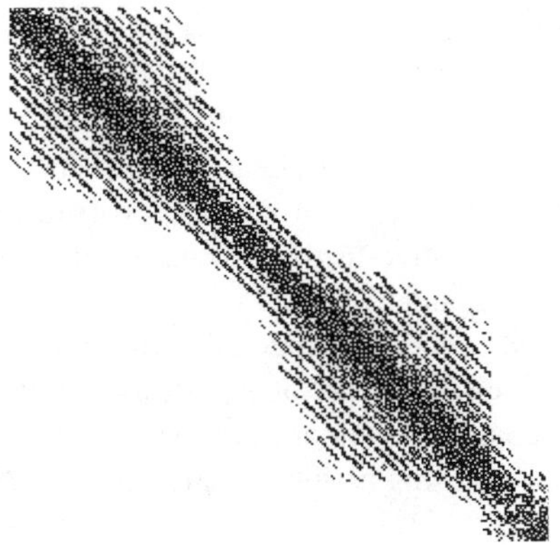

Fig. 3.8. Population of influence matrix $\mathbf{M}^\infty$ for zero-element thresholds $\varepsilon_z = 10^{-4}$

Fig. 3.9. Population of influence matrix $\mathbf{M}^\infty$ for zero-element thresholds $\varepsilon_z = 10^{-5}$

Fig. 3.10. Population of influence matrix $\mathbf{M}^\infty$ for zero-element thresholds $\varepsilon_z = 10^{-6}$

3.8 Application of hierarchical matrices

In this work, the concept of hierarchical or $\mathcal{H}$-matrices [29, 30, 46, 64, 65, 66, 64, 128] is applied to matrix vector multiplication with fully populated matrices derived from SBFEM formulation. $\mathcal{H}$-matrices are evolved from the panel clustering [67] which provides an alternative to the fast multipole method [62]. This method reduces the complexity of standard matrix operations from a quadratic to a quasi-linear ($O(n\log^{\alpha} n)$, $\alpha \in \mathbb{R}$) cost in storage and computation time. The principle of hierarchical matrices is based on hierarchical partitioning of a given matrix into block matrices of smaller dimension, which are approximated by so-called low rank matrices. The implementation of this method is strongly facilitated by using the HLIB [28].

3.8.1 Low rank matrices

Low rank matrices represent an important class of *data sparse* matrices [128], which enable an optimal storage and matrix arithmetic. The main idea of a low rank representation is that a matrix $\mathbf{A} \in \mathbb{R}^{n \times n}$ with $\text{rank}(A) = r << n$ can be written as:

$$\mathbf{A}_{rank\,r} = \sum_{k=1}^{r} \mathbf{a}_k \mathbf{b}_k^T \quad \mathbf{a}_k, \mathbf{b}_k^T \in \mathbb{R}^n \ . \tag{3.139}$$

Thus, matrix $\mathbf{A}$ can be represented by $2r$ vectors $\mathbf{a}_k$ and $\mathbf{b}_k$, which results in a memory requirement (mem) to store matrix $\mathbf{A}$:

$$\text{mem}(\mathbf{A}_{\text{rank}\,r}) = 2rn \ . \tag{3.140}$$

The multiplication of a rank r matrix $\mathbf{A}$ with a given vector $\mathbf{x} \in \mathbb{R}^n$ is then:

$$\mathbf{A}_{\text{rank}\,r}\mathbf{x} = \sum_{k=1}^{r} \mathbf{a}_k \mathbf{b}_k^T \mathbf{x} = \sum_{k=1}^{r} (\mathbf{b}_k^T \mathbf{x})\mathbf{a}_k \ . \tag{3.141}$$

One can assert that matrices $\mathbf{A}$ with a low rank $r << n$ can be efficiently stored and used. In general, an exact low rank matrix representation of original matrix $\mathbf{A}$ is not possible. Thus, low rank matrix approximation is introduced in subsection 3.8.3.

3.8.2 Hierarchical partitioning

The basis of hierarchical matrices is a partitioning of a matrix $\mathbf{A}$ into block matrices $\mathbf{A}_{ij}$ and a low rank representation of those block matrices [128]. Every random entry $A[l,k]$ of matrix $\mathbf{A} \in \mathbb{R}^{n \times n}$ is identified by the entry $A_{ij}[l_j,k_i]$ of a block matrix $\mathbf{A}_{ij} \in \mathbb{R}^{n_j \times n_i}$, if a unique mapping between indices $k_i \leftrightarrow k$ and $l_j \leftrightarrow l$ exists. With $k_i = 1,\ldots,n_i$ and $l_j = 1,\ldots,n_j$ the related indices k and l belong to index subsets I_i and I_j of the original index set $I = \{1,\ldots,n\}$. With these index subsets $I_i, I_j \subset I$ it is possible to describe a block matrix $\mathbf{A}_{ij} \in \mathbb{R}^{n_j \times n_i}$. The significant advantage of this description is that the partitioning of a matrix $\mathbf{A}$ corresponds to the partitioning of an index set I.

A *partitioning* of index set I is

$$P(I) = \{I_1, \ldots, I_p\} = I_{i\,i=1}^{\,p}, \quad p \in \mathbb{N} \tag{3.142}$$

with pairwise disjoint index sets I_i of dimension $n_i = \dim(I_i)$ with $I_i \cap I_j = \emptyset$ for $i \neq j$. The partitioning $P(I)$ is a complete decomposition of I, thus

$$I = \bigcup_{i=1}^{p} I_i, \quad \sum_{i=1}^{p} n_i = n \,. \tag{3.143}$$

To obtain a *hierarchical partitioning* with a given level λ, an initial partitioning has to be determined followed by a recursive algorithm, which has to be repeated at each level. The initial partitioning is given by

$$P^0(I) = \{I_i^0\}_{i=1}^{p_0} = \{I\}, \quad p_0 = 1 \,. \tag{3.144}$$

Then, for level $\lambda = 0, 1, \ldots, L$ recursive partitioning can be written as:

$$P^\lambda(I) = \{I_i^\lambda\}_{i=1}^{p_\lambda} \,, \tag{3.145}$$

where I_i^λ are pairwise disjoint index sets of equal level and dimension $n_i^\lambda = \dim(I_i^\lambda)$ and $I_i^\lambda \cap I_j^\lambda = \emptyset$ for $i \neq j$. One can sum up these steps into a *complete decomposition* of the index set I for each level λ:

$$I = \bigcup_{i=1}^{p_\lambda} I_i^\lambda, \quad \sum_{i=1}^{p_\lambda} n_i^\lambda = n \,. \tag{3.146}$$

This kind of partitioning is *hierarchical*, because all index sets I_i^λ build a hierarchy: for all $i = 1, \ldots, p_\lambda$ and levels $\lambda = 1, \ldots, L$ only one index set $I_j^{\lambda-1}$ exists, with $I_i^\lambda \subset I_j^{\lambda-1}$. Without loss of generality it is assumed that index set I_i^λ consists of connected indices. Hence, I_i^λ build a *cluster tree T* (see Fig.3.11).

With hierarchical partitioning of index set I a hierarchical partitioning of index set $I \times I$ can be defined:

$$P_{\mathcal{H}}(I, T) = \{I_i^\lambda \times I_j^\lambda : I_i^\lambda, I_j^\lambda \in P^\lambda(I), \lambda = 0, \ldots, L\} \,. \tag{3.147}$$

Bisection method

A suitable partitioning method represents the *bisection method* [128], which belongs to the class of geometrical partitioning.

The centroid $\mathbf{s}$ of a point set $\{\mathbf{x}_k\}_{k=1}^n$ is defined by:

$$\mathbf{s} := \frac{1}{n} \sum_{k=1}^{n} \mathbf{x}_k \,. \tag{3.148}$$

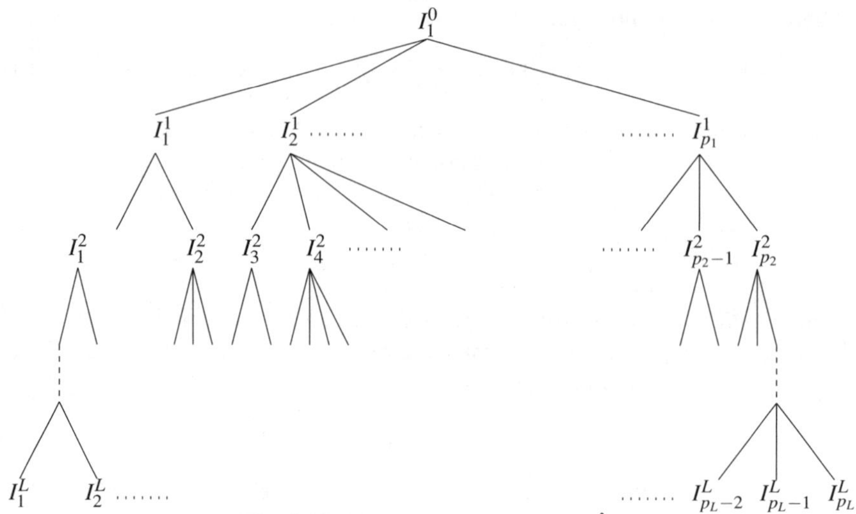

Fig. 3.11. Cluster tree of index sets I_i^{λ}

To find the principal direction $\mathbf{w}$, defined as normal vector through centroid $\mathbf{s}$, one has to solve the maximum problem [46]:

$$\mathbf{w} := \arg \max_{\mathbf{v} \in \mathbb{R}^d,\, \|\mathbf{v}\|_2 = 1} \sum_{k=1}^{n} (\mathbf{v}^T (\mathbf{x}_k - \mathbf{s}))^2 , \tag{3.149}$$

which divides the index set into two parts:

$$I_1 := \{ i \in I \,|\, (\mathbf{w}^T (\mathbf{x}_i - \mathbf{s})) \geq 0 \} \text{ and} \tag{3.150}$$

$$I_2 := \{ i \in I \,|\, (\mathbf{w}^T (\mathbf{x}_i - \mathbf{s})) < 0 \} . \tag{3.151}$$

The principle direction $\mathbf{w} \in \mathbb{R}^d$ can be obtained by solving the general eigenvalue problem:

$$\mathbf{K}\mathbf{w} = \lambda_{max}\mathbf{w} , \tag{3.152}$$

with $\mathbf{K}$ being the covariant matrix:

$$\mathbf{K} = \sum_{k=1}^{n} (\mathbf{x}_k - \mathbf{s})(\mathbf{x}_k - \mathbf{s})^T , \quad \mathbf{K} \in \mathbb{R}^{d \times d} . \tag{3.153}$$

The resulting algorithm for partitioning a point set $\{\mathbf{x}_k\}_{k=1}^{n}$ with the bisection method follows:

1. Compute the centroid $\mathbf{s}$ of index set $\{\mathbf{x}_k\}_{k=1}^{n}$.
2. Build covariant matrix $\mathbf{K}$.

3. Find largest eigenvalue $\lambda_{max}(\mathbf{K})$ and its corresponding eigenvector $\mathbf{w}$ (principal direction).
4. With the principal direction $\mathbf{w}$ divide index set I into subsets I_1 and I_2.
5. Apply algorithm recursively until given dimension L is reached.

The application of the proposed algorithm results in a cluster tree of index sets I_j^λ with

$$I_1^0 = 0, \quad I_1^{\lambda-1} = I_{2i-1}^\lambda \cup I_2^\lambda; \quad \text{with } i = 1,\ldots,2^{\lambda-1}, \lambda = 1,\ldots,L. \tag{3.154}$$

Algorithm for hierarchical partitioning

In the following, an algorithm for hierarchical partitioning of a matrix $\mathbf{A}$ is given [128]. It is started with level $\lambda = 0$ and will be repeated for each level:

1. Check, if block matrices $\mathbf{A}_{ij}^\lambda$ are r-admissible with the reduced singular value decomposition (rSVD)[2].
2. If the block matrix $\mathbf{A}_{ij}^\lambda$ is r-admissible, describe the matrix with its low rank representation.
3. If the block matrix is not r-admissible, there are two possibilities:
   a) $\lambda < L$ (highest level not reached): The block matrix $\mathbf{A}_{ij}^\lambda$ will be partitioned with the bisection method for all children $I_k^{\lambda+1}$ and $I_l^{\lambda+1}$ of the original index sets I_i^λ and I_j^λ. All block matrices $\mathbf{A}_{kl}^{\lambda+1}$ at next level $\lambda+1$ will be built. These matrices will be tested on r-admissibility according to step 1.
   b) $\lambda = L$ (highest level reached): At leaf level arrived. The block matrix $\mathbf{A}_{ij}^L$ cannot be represented as a low rank matrix. It is represented as a dense matrix with $n_i^L \cdot n_j^L$ entries.

r-Admissibility

Elements of the last level of a hierarchical partitioning are called *leafs* of the cluster tree . Leafs will not be partitioned further. During construction of a cluster tree, it is necessary to have a criterion for elements, whether they are leafs or whether they should be further partitioned. This criterion is the admissibility. A block matrix $\mathbf{A}_{ij}^\lambda \in \mathbb{R}^{n_j \times n_i}$ is admissible and the associated index sets I_i^λ and I_j^λ are admissible to each other, respectively, if the block matrix $\mathbf{A}_{ij}^\lambda$ allows a representation as a low rank or rank r matrix. Admissible block matrices are called r-admissible, if a rank r representation is possible. Inadmissible block matrices are partitioned until the last level $\lambda = L$ is reached. The r-admissibility of a block matrix can be tested by reduced singular value decomposition (see next paragraph).

---

[2] For r-admissibility and rSVD refer to next paragraphs

Reduced singular value decomposition

A *reduced singular value decomposition* (rSVD) of a matrix $\mathbf{M}$ is a factorisation of the form

$$\mathbf{M} = \mathbf{U}\mathbf{S}\mathbf{V}^T \tag{3.155}$$

with matrices $\mathbf{U} \in \mathbb{R}^{n \times k}, \mathbf{V} \in \mathbb{R}^{m \times k}$ that have orthonormal columns and a diagonal matrix $\mathbf{S} \in \mathbb{R}^{k \times k}$, see Fig. 3.12, where the diagonal entries are $S_{11} \geq S_{22} \geq \ldots \geq S_{kk} > 0$. The diagonal entries of $\mathbf{S}$ are called the *singular values* of $\mathbf{M}$.

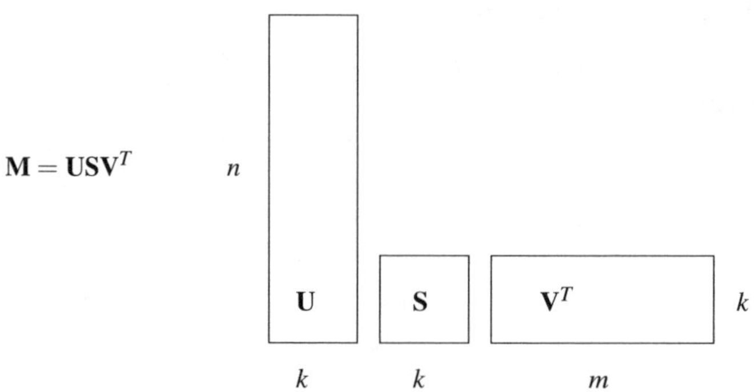

Fig. 3.12. Factorisation of $\mathbf{M}$ with rSVD

A SVD of a general matrix can be computed by a standard LAPACK subroutine [122] with complexity $O(\min(n,m)\max(n,m)^2)$. It is possible to compute a rSVD with only $O(k^2 \max(n,m))$ complexity [30]. Let $\mathbf{M} = \mathbf{a}_k \mathbf{b}_k^T$ be a matrix in $\mathcal{H}$-matrix representation, then a rSVD $\mathbf{M} = \mathbf{U}\mathbf{S}\mathbf{V}^T$ can be computed with the three following steps:

1. Compute reduced QR-factorisations of $\mathbf{a}_k$ and $\mathbf{b}_k$: $\mathbf{a}_k = \mathbf{Q}_A \mathbf{R}_A$, $\mathbf{b}_k = \mathbf{Q}_B \mathbf{R}_B$ with matrices $\mathbf{Q}_A \in \mathbb{R}^{n \times k}$, $\mathbf{Q}_B \in \mathbb{R}^{m \times k}$, $\mathbf{R}_A \in \mathbb{R}^{k \times k}$, $\mathbf{R}_B \in \mathbb{R}^{k \times k}$ with complexity $O((n+m)k^2)$.
2. Compute a SVD of $\mathbf{R}_A \mathbf{R}_B^T = \mathbf{U}'\mathbf{S}\mathbf{V}'$ with complexity $O(k^3)$.
3. Finally, compute $\mathbf{U} = \mathbf{Q}_A \mathbf{U}'$ and $\mathbf{V} = \mathbf{Q}_B \mathbf{V}'$ with complexity $O((n+m)k^2)$.

3.8.3 Approximation with low rank matrices

Because an exact low rank representation of a matrix is not possible in general, an approximation of the matrix vector product with low rank matrices is introduced here.

A block matrix $\mathbf{A}_{ij}^{\lambda} \in \mathbb{R}^{n_j^{\lambda} \times n_i^{\lambda}}$ is (r,ε)-admissible, if it is possible to approximate $\mathbf{A}_{ij}^{\lambda}$ with a matrix $\mathbf{A}_{ij;r}^{\lambda}$ of same dimension and rank $\mathbf{A}_{ij;r}^{\lambda} \leq r$ with given error tolerance ε. In the following, symmetric matrices are considered, because influence matrices $\mathbf{M}^{\infty}$ of the SBFEM are symmetric.

A given symmetric matrix $\mathbf{A} = \mathbf{A}^{T} \in \mathbb{R}^{n \times n}$ with n non-negative real eigenvalues $\lambda_1, \ldots, \lambda_n$ has related n eigenvectors $\{\mathbf{v}_k\}_{k=1}^{n}$, which build an orthonormal system with respect to the Euklidian scalar product. The matrix $\mathbf{V}$ built from these eigenvectors is orthogonal ($\mathbf{V}^T \mathbf{V} = \mathbf{V}\mathbf{V}^T = \mathbf{I}$). Furthermore, the diagonal matrix $\mathbf{D}$ of eigenvalues is defined as:

$$\mathbf{D} = \mathrm{diag}(\lambda_k(\mathbf{A}))_{k=1}^{n} \,. \tag{3.156}$$

A symmetric matrix $\mathbf{A}_r = \mathbf{A}_r^{T} \in \mathbb{R}^{n \times n}$ with rank $\mathbf{A}_r \leq r < n$ is sought as solution of the minimisation problem

$$\|\mathbf{A} - \mathbf{A}_r\|_2 = \min_{\mathbf{B}=\mathbf{B}^T \in \mathbb{R}^{n \times n},\, \mathrm{rank}\mathbf{B} \leq r} \|\mathbf{A} - \mathbf{B}\|_2 \,. \tag{3.157}$$

The rank r approximation $\mathbf{A}_r$ of $\mathbf{A}$ has a maximum of r non-zero eigenvalues $\lambda_k(\mathbf{A}_r)$. With the well-known factorisation

$$\mathbf{A} = \mathbf{V}\mathbf{D}\mathbf{V}^T = \sum_{k=1}^{n} \lambda_k(\mathbf{A})\mathbf{v}_k\mathbf{v}_k^T \tag{3.158}$$

the approximation $\mathbf{A}_r$ is defined as:

$$\mathbf{A}_r = \mathbf{V}\mathbf{D}_r\mathbf{V}^T = \sum_{k=1}^{r} \lambda_k(\mathbf{A})\mathbf{v}_k\mathbf{v}_k^T \tag{3.159}$$

with diagonal matrix

$$\mathbf{D}_r = \begin{pmatrix} \lambda_1(\mathbf{A}) & & & & & \\ & \ddots & & & & \\ & & \lambda_r(\mathbf{A}) & & & \\ & & & 0 & & \\ & & & & \ddots & \\ & & & & & 0 \end{pmatrix} \,. \tag{3.160}$$

The solution of the minimisation problem (3.157) is the low rank matrix representation (3.159), and

$$\min_{\mathbf{B}=\mathbf{B}^T \in \mathbb{R}^{n \times n},\, \mathrm{rank}\mathbf{B} \leq r} \|\mathbf{A} - \mathbf{B}\|_2 = \|\mathbf{A} - \mathbf{A}_r\|_2 = \lambda_{r+1}(\mathbf{A}) \,, \tag{3.161}$$

where $\|\cdot\|_2$ is the Euklidian matrix norm. An equivalent minimisation problem can also be formulated for the Frobenius matrix norm $\|\cdot\|_F$ with the solution:

$$\min_{\mathbf{B}=\mathbf{B}^T \in \mathbb{R}^{n \times n},\, \mathrm{rank}\mathbf{B} \leq r} \|\mathbf{A} - \mathbf{B}\|_F = \|\mathbf{A} - \mathbf{A}_r\|_F = \sqrt{\sum_{k=r+1}^{n} \lambda_k(\mathbf{A})^2} \,. \tag{3.162}$$

For a given vector $\mathbf{x} \in \mathbb{R}^n$ the exact matrix vector product $\mathbf{z} = \mathbf{A}\mathbf{x}$ is replaced by the approximated product $\tilde{\mathbf{z}} = \mathbf{A}_r\mathbf{x}$. An error estimation can be done with help of the Euklidian matrix norm:

$$\|\mathbf{z} - \tilde{\mathbf{z}}\|_2 = \|(\mathbf{A} - \mathbf{A}_r)\mathbf{x}\|_2 \leq \|\mathbf{A} - \mathbf{A}_r\|_2 \|\mathbf{x}\|_2 = \lambda_{r+1}(\mathbf{A}) \|\mathbf{x}\|_2 \,, \tag{3.163}$$

or with the Frobenius norm:

$$\|\mathbf{z} - \tilde{\mathbf{z}}\|_2 = \|(\mathbf{A} - \mathbf{A}_r)\mathbf{x}\|_2 \leq \|\mathbf{A} - \mathbf{A}_r\|_F \|\mathbf{x}\|_2 = \sqrt{\sum_{k=r+1}^{n} \lambda_k(\mathbf{A})^2} \, \|\mathbf{x}\|_2 \,. \tag{3.164}$$

Error estimation formula (3.164) is weaker than (3.163), but computationally less expensive.

3.8.4 Matrix vector multiplication

The adaptive arithmetic of hierarchical matrices allows an efficient matrix vector multiplication also for fully populated matrices [29, 92]. With

- the mapping of global index set I to local index set: I_i^λ $R_i^\lambda : I \to I_i^\lambda$; $R_i^\lambda \in \mathbb{R}^{n_i^\lambda \times n}$,
- the corresponding hierarchical partitioning of index set $I \times I$: $P_{\mathcal{H}}(I,T)$, and
- the partitioning of all admissible index pairs $(I_i^\lambda, I_j^\lambda)$: $P_{\mathcal{H}}^Z(I,T) \subset P_{\mathcal{H}}(I,T)$,

a representation for a hierarchical matrix $\mathbf{A}$ is [128, 46]:

$$\mathbf{A} = \sum_{(I_i^\lambda, I_j^\lambda) \in P_{\mathcal{H}}^Z(I,T)} (\mathbf{R}_j^\lambda)^T \mathbf{A}_{ij}^\lambda \mathbf{R}_i^\lambda \,, \tag{3.165}$$

with rank r block matrices

$$\mathbf{A}_{ij}^\lambda = \sum_{k=1}^{r} \mathbf{a}_{j;\lambda;k} \mathbf{b}_{i;\lambda;k}^T \,. \tag{3.166}$$

The matrix vector product $\mathbf{z} = \mathbf{A}\mathbf{x}$ can be written with (3.165) as:

$$\mathbf{z} = \mathbf{A}\mathbf{x} = \sum_{(I_i^\lambda, I_j^\lambda) \in P_{\mathcal{H}}^Z(I,T)} (\mathbf{R}_j^\lambda)^T \mathbf{A}_{ij}^\lambda \mathbf{R}_i^\lambda \mathbf{x} = \sum_{(I_i^\lambda, I_j^\lambda) \in P_{\mathcal{H}}^Z(I,T)} (\mathbf{R}_j^\lambda)^T \mathbf{z}_{ij;\lambda} \tag{3.167}$$

with

$$\mathbf{z}_{ij;\lambda} = \mathbf{A}_{ij}^\lambda \mathbf{x}_{i,\lambda}, \quad \mathbf{x}_{i,\lambda} = \mathbf{R}_i^\lambda \mathbf{x} \,. \tag{3.168}$$

Thus, the matrix vector multiplication is reduced to local matrix vector multiplication

$$\mathbf{z}_{ij;\lambda} = \mathbf{A}_{ij}^\lambda \mathbf{x}_{i;\lambda} = \left(\sum_{k=1}^{r} \mathbf{a}_{j;\lambda;k} \mathbf{b}_{i;\lambda;k}^T \right) \mathbf{x}_{i;\lambda} = \sum_{k=1}^{r} \left(\mathbf{b}_{i;\lambda;k}^T \mathbf{x}_{i;\lambda} \right) \mathbf{a}_{j;\lambda;k} \,, \tag{3.169}$$

and assemblage of solution vectors $\mathbf{z}_{ij;\lambda}$. The resulting algorithm for matrix vector multiplication follows:

Loop over all admissible index pairs $(I_i^\lambda, I_j^\lambda) \in P_{\mathcal{H}}^Z(I,T)$.

1. Determine local vectors $\mathbf{x}_{i;\lambda} = \mathbf{R}_i^\lambda \mathbf{x}$.

2. Compute local matrix vector multiplications $\mathbf{z}_{ij;\lambda} = \sum\limits_{k=1}^{r} \left(\mathbf{b}_{i;\lambda;k}^T \mathbf{x}_{i;\lambda} \right) \mathbf{a}_{j;\lambda;k}$ with complexity $r(n_i^\lambda + n_j^\lambda)$.

3. Assemblage of local parts $(\mathbf{R}_j^\lambda)^T \mathbf{z}_{ij;\lambda}$ to solution vector $\mathbf{z}$.

The total complexity of matrix vector multiplication $\mathbf{z} = \mathbf{A}\mathbf{x}$ is:

$$O(\mathbf{A}\mathbf{x}) = r \sum_{(I_i^\lambda, I_j^\lambda) \in P_{\mathcal{H}}^Z(I,T)} (n_i^\lambda + n_j^\lambda) . \tag{3.170}$$

The total complexity (equivalent to total computational effort) depends on the structure of $P_{\mathcal{H}}^Z(I,T)$ and therefore on the used admissibility condition.

4. Benchmark examples

In this section, four two- and three-dimensional examples of soil-structure interaction
are presented. The coupling of finite and boundary elements, e.g., for soil-structure
interaction or geotechnical problems is successfully applied [16, 115, 63]. For the
presented examples in this section, results from a BE or coupled FE/BE procedure
exist. Therefore, results of the presented coupled FE/SBFE application can be analysed concerning accuracy.

4.1 Two-dimensional benchmark examples of soil-structure interaction in the time domain

To demonstrate the precision and applicability of a coupled FE/SBFE procedure with
effective modifications three two-dimensional numerical examples are presented in
this subsection. These examples have been studied by von Estorff and Prabucki [51],
and Stamos et al. [127], respectively, as a benchmark for their coupled FE/BE formulation. Therefore they can easily be compared to the presented method.

4.1.1 Semi-infinite rectangular domain with vertical load

Purpose of the first study is to compare results obtained by the coupled FE/SBFE
approach of a semi-infinite rectangular domain with a corresponding coupled FE/BE
solution of Estorff and Prabucki [51].

Fig. 4.1 illustrates the elastic region [1] with applied load, load function and location of observation point A. The elastic region (plane strain) has a Young's modulus
$E = 266MPa$, Poisson's ration $v = 0.33$ and a density $\rho = 2000\frac{kg}{m^3}$.

The discretisation of this problem with linear finite and boundary elements, and
linear scaled boundary finite elements is shown in Figs. 4.2 and 4.3, respectively.
For the coupled FE/BE discretisation 24 finite elements and 14 boundary elements
are used which is identical to the discretisation by Estorff and Prabucki [51] in order to gain comparable results. For the semi-infinite domain with parallel sides, the
scaling centre for the scaled boundary transformation has to be placed in infinite
distance to the FE/SBFE interface. Note that no discretisation of the side faces of

---

[1] In this work, linear elastic material behaviour is assumed. For non-linear simulations dynamic soil parameters could be used.

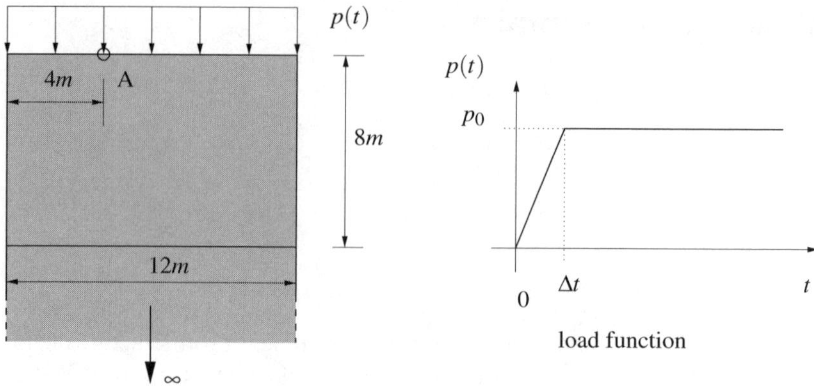

Fig. 4.1. Semi-infinite rectangular domain: system and load function

the infinite strip is required, when using a scaled boundary finite element mesh at the near-field/far-field interface of the infinite strip, see Fig. 4.3. The drawback of necessarily placed boundary elements along the semi-infinite strip (here 4 boundary elements are placed on each side, can be reduced by using *infinite boundary elements* [17, 33].

A time-step size $\Delta t = 0.002s$ is chosen for all benchmark examples. The horizontal and vertical displacements of point A are studied.

Fig. 4.4 shows the displacement of point A, for a time interval $0 \leq t \leq 0.14s$, calculated with a coupled FE/BE and a coupled FE/SBFE methodology. Both methods generate the same horizontal displacement, as well as vertical displacement of the observation point A.

Fig. 4.5 depicts results of a coupled FE/SBFE approach for a longer simulation time $0 \leq t \leq 1s$. A rigorous "conventional" calculation is compared with the presented recursive algorithm. The results show excellent agreement of rigorous and recursive algorithm, no differences can be observed.

The entries of the influence matrices grow linear from a certain time t_m, see Sec. 3.7.1. This time-station t_m is determined as $t_m = 0.16s$ (corresponds to linear behaviour after $m = 80$ time-steps). According to (3.136) and (3.137), the computational effort is reduced to 0.294, compared to the rigorous algorithm, so the reduction of necessary operations (equivalent to CPU time consumption) is 70.6%.

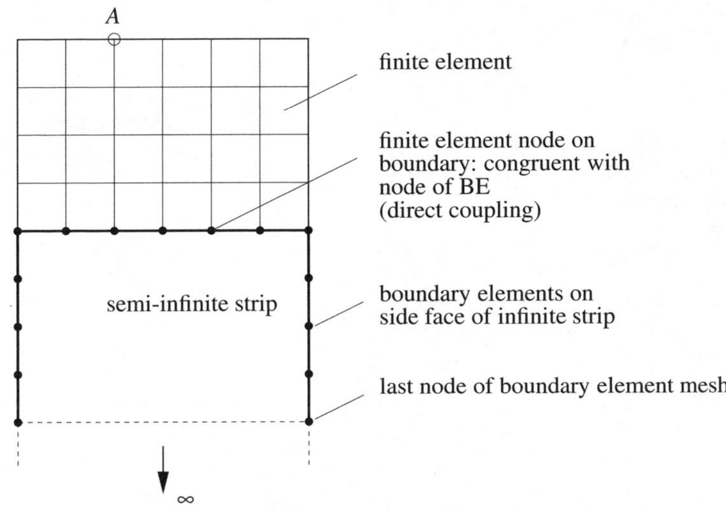

Fig. 4.2. Coupled FE/BE discretisation of semi-infinite rectangular domain

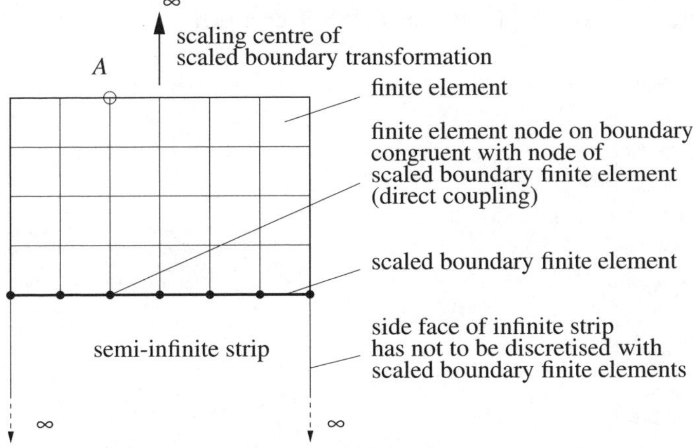

Fig. 4.3. Coupled FE/SBFE discretisation of semi-infinite rectangular domain

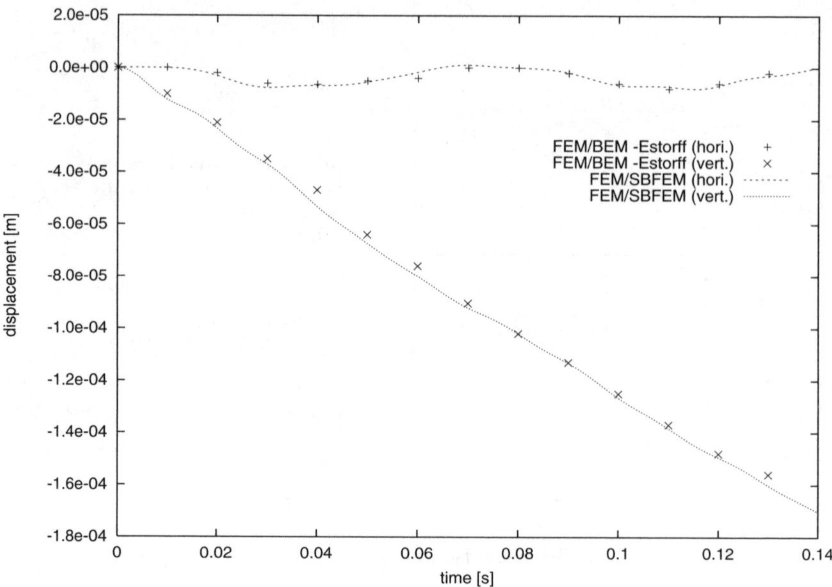

Fig. 4.4. Semi-infinite rectangular domain: horizontal and vertical displacement of point *A* due to a vertical load: coupled FE/BE and FE/SBFE approach

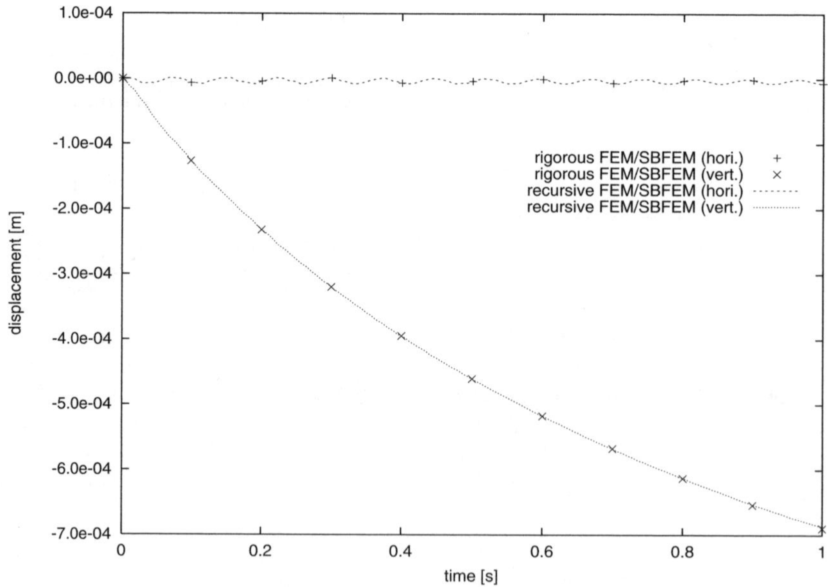

Fig. 4.5. Semi-infinite rectangular domain: horizontal and vertical displacement of point *A* due to a vertical load: coupled FE/SBFE approach, rigorous and recursive algorithm

4.1.2 Response of an elastic half-space

For the second example of this subsection, an infinite half-space with a time-dependent horizontal load is considered which was previously examined by von Estorff and Prabucki [51]. Fig. 4.6 shows the applied load and two observation points (A and B) on the surface of the elastic half-space. The material properties are the same as for the first example. Fig. 4.7 shows the time-dependent horizontal load $p(t)$, called Ricker-wavelet , where $p(t)$ is given by:

$$p(t) = (1 - 2\tau^2)\exp(-\tau^2) \quad \text{with } \tau = t\pi - 3 \ . \tag{4.1}$$

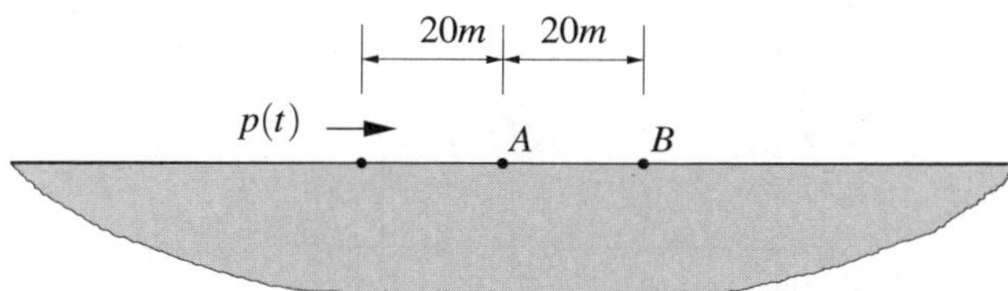

Fig. 4.6. Elastic half-space: location of load and observation point

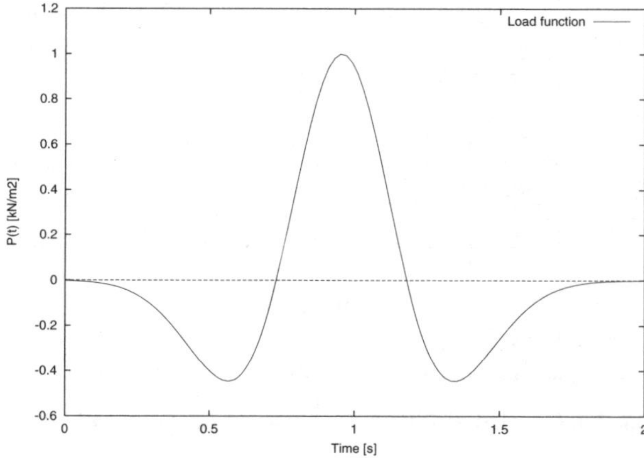

Fig. 4.7. Load function $p(t)$ (Ricker wavelet)

The discretisation, again of a coupled FE/BE and a FE/SBFE system is shown in Figs. 4.8 and 4.9, respectively. The same discretisation as used by von Estorff and Prabucki [51] was constructed in order to achieve comparable results. The scaling centre for the scaled boundary approach lies inside the FE domain, on the surface of the infinite half-space. For the coupled FE/SBFE approach only the scaled boundary finite elements on the near-field/far-field interface are necessary, while for the coupled FE/BE discretisation a few additional boundary elements on the surface of the half-space on both sides of the FE mesh are required.

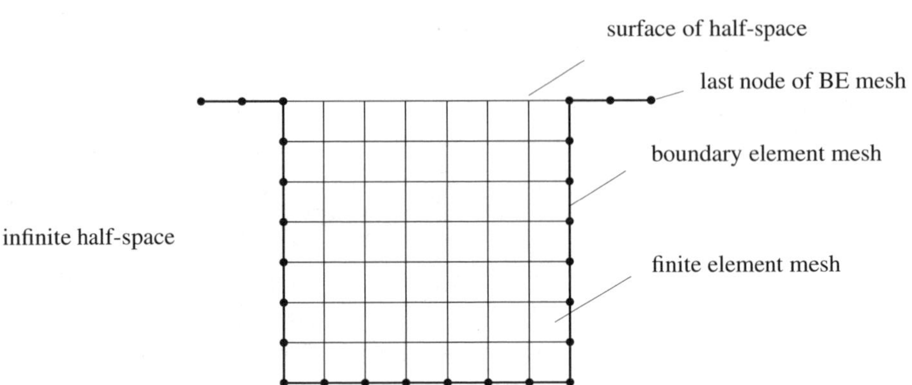

Fig. 4.8. Coupled FE/BE discretisation of elastic half-space

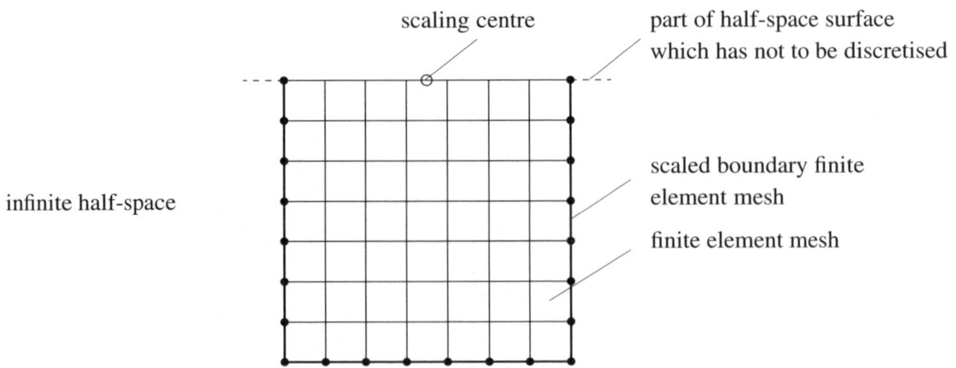

Fig. 4.9. Coupled FE/SBFE discretisation of elastic half-space

Results of the calculation (horizontal displacements of points A and B) are shown for both approaches in Figs. 4.10 and 4.11, respectively. Again, neither major deviations can be observed between the coupled FE/BE and FE/SBFE approach, nor between the direct and recursive algorithm. For this example, the simulation is carried out for $t \leq 4s$. For the recursive algorithm $m = 309$ is found, so with $\Delta t = 0.002s$ ($n = 2000$) the computational effort is reduced to 0.285 compared to the rigorous algorithm, thus the reduction of necessary operations (equivalent to CPU time consumption) is 71.5%.

For both examples, the reduction of CPU time consumption due to the recursive algorithm is huge, while nearly no loss of accuracy can be observed.

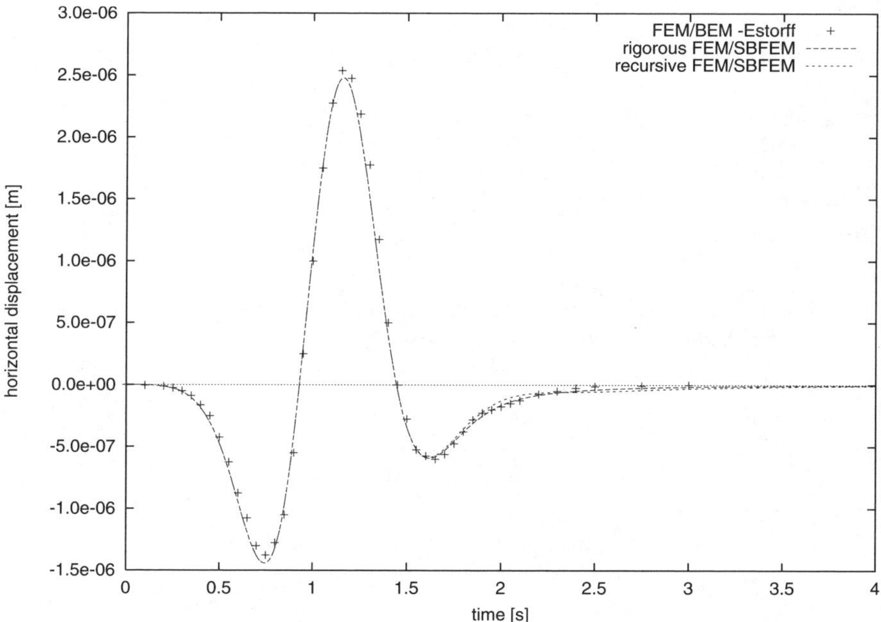

Fig. 4.10. Elastic half-space: horizontal displacement of point A due to a horizontal load - coupled FE/SBFE approach, rigorous and recursive algorithm

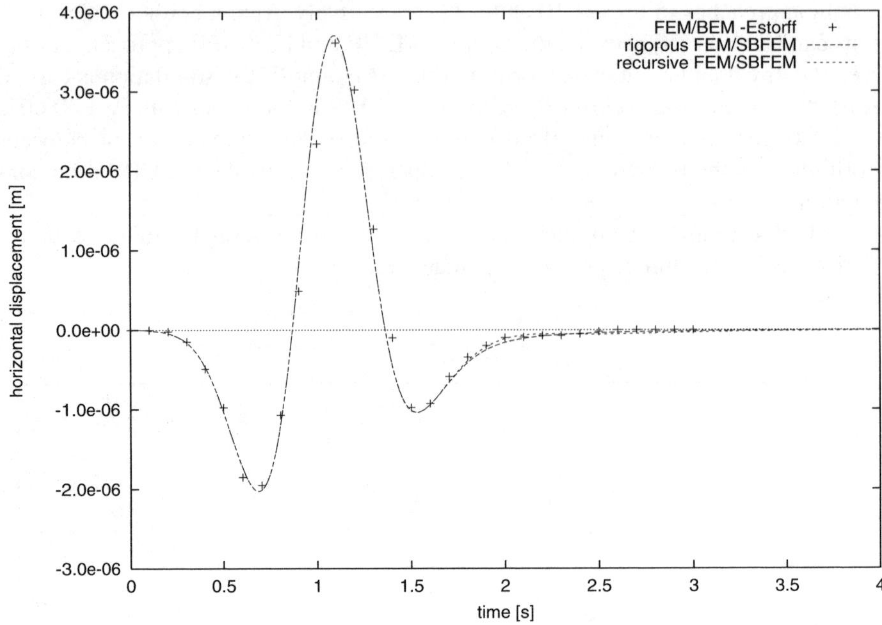

Fig. 4.11. Elastic half-space: horizontal displacement of point B due to a horizontal load - coupled FE/SBFE approach, rigorous and recursive algorithm

4.1.3 Road-tunnel traffic system

In this subsection, a two-dimensional road-tunnel traffic system is considered. This system was already analysed by Stamos et al. [127] with a coupled FE/BE procedure, and thus results gained by the proposed coupled FE/SBFE procedure can be easily compared. In this example, dynamic interaction effects in tunnel systems subjected to above and below ground traffic loads are studied numerically under conditions of plane strain. Linear elastic material behaviour for both the structure and the soil is assumed.

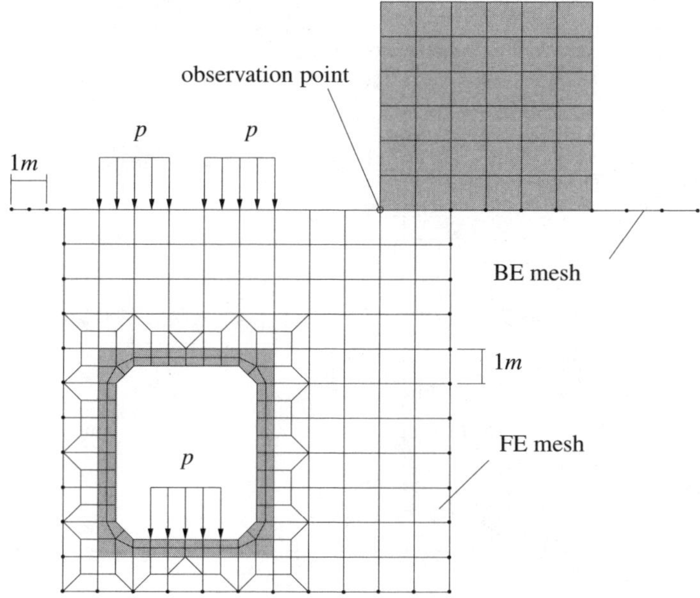

Fig. 4.12. Coupled FE/BE discretisation and loadings of road-tunnel traffic system [127]

Fig. 4.12 depicts the discretisation scheme for the coupled FE/BE method, with isoparametric four-node (linear) plane strain finite elements and two-node boundary elements used by Stamos et al. [127]. The grey marked tunnel of Fig. 4.12 has a Young's modulus of $30000MPa$, a Poisson's ratio $\nu = 0.33$ and a density ρ of $2000\frac{kg}{m^3}$, the also grey marked building in the neighbourhood of the traffic induced loadings has a Young's modulus of $3000MPa$, Poisson's ratio $\nu = 0.3$ and a density $\rho = 2000\frac{kg}{m^3}$, and finally, the surrounding soil, which is assumed to behave linear elastic, has a Young's modulus of $266MPa$, a Poisson's ratio $\nu = 0.33$ and a density ρ of $2000\frac{kg}{m^3}$.[2]

The discretisation of the same road-tunnel system, but now for a coupled FE/SBFE approach is shown in Fig. 4.13. Finite elements of same shape, size and type

---

[2] Material parameters chosen according to [127].

(isoparamteric four-node plane strain) were chosen, but additional elements were placed under the building. For a scaled boundary discretisation of the infinite half-space, represented by two-node (linear) plane strain line elements, no elements have to be placed on the surface. A refined version of the mesh is depicted in Fig. 4.14, where the numerical model is generated by linear finite and scaled boundary elements as well. Here, the default element size is reduced from $1m \times 1m$ (coarse mesh of Fig. 4.13) to $0.5m \times 0.5m$.

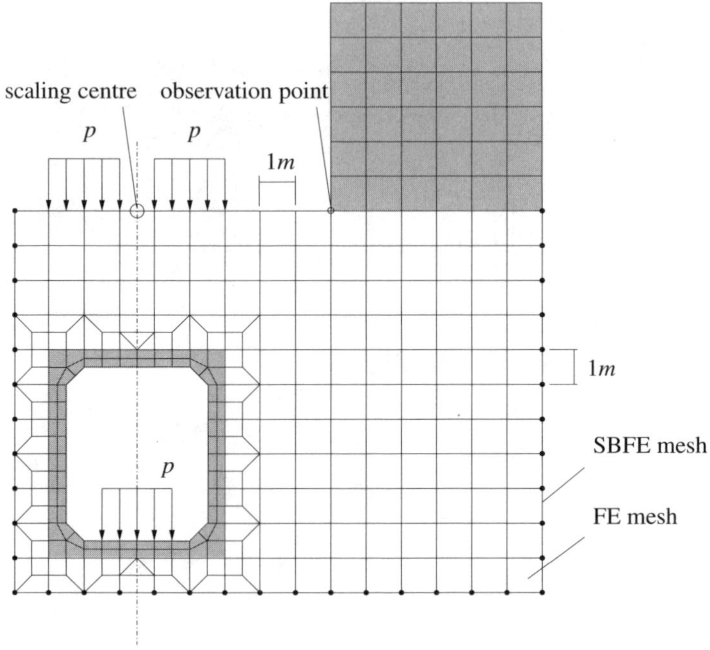

Fig. 4.13. Coarse FE/SBFE discretisation of road-tunnel traffic system

For all traffic induced (vertical) loads $p(t)$ a Heaviside step function (see Fig. 4.15) of $1kPa$ for $0 \le t \le 0.02s$, and 0 for $t > 0.02s$ is applied, see Figs. 4.12, 4.13 and 4.14. A time-step $\Delta t = 2.5 \cdot 10^{-4}s$ is chosen for the simulation, with a simulation time of $0.16s$, the total number of time-steps is 640.

The simulation results for $t \le 0.16s$ are displayed in Figs. 4.16 and 4.17. The calculated vertical displacement of the FE/SBFE procedure is compared to a coupled FE/BE method. The results are in good agreement, where results by the recursive algorithm are close to those of the rigorous calculation. Calculated vertical displacements, gained by a rigorous FE/SBFE simulation with refined discretisation are in excellent agreement with FE/BE and FE/SBFE (coarse mesh) results.

For the recursive algorithm, the time-station t_m is determined as $t_m = 0.047s$. This corresponds to a linear behaviour of the influence matrices after $m = 188$ time-

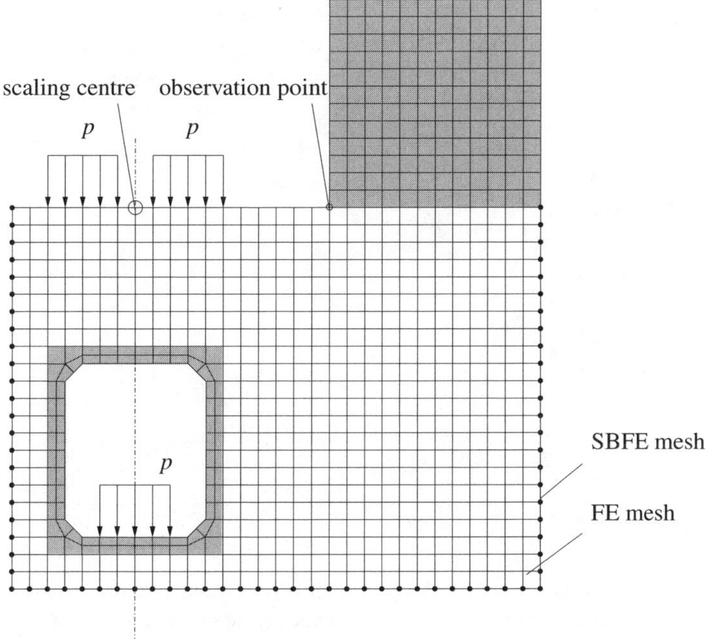

Fig. 4.14. Refined FE/SBFE discretisation of road-tunnel traffic system

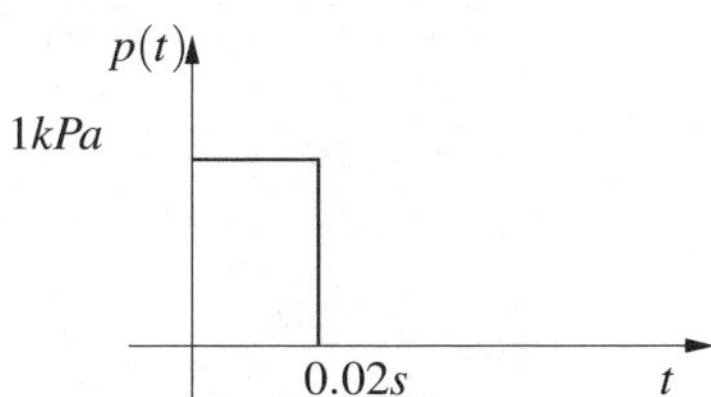

Fig. 4.15. Time-dependency of the vertical load $p(t)$

steps. According to (3.136) and (3.137), the computational effort is reduced to 0.50 ($\frac{m}{n} = 0.294$, with total number of time-steps $n = 640$), compared to the rigorous algorithm, so a 50.0% reduction of necessary operations is achieved.

These two-dimensional examples show that the computational effort can be dramatically reduced to an acceptable level with marginal loss of accuracy by using the presented recursive algorithm of the SBFEM. This algorithm is based on an approximation of the unit-impulse response matrices by linear time dependence after a certain time-step. Further approximations in space lead to additional reductions in computational costs, where a sparse structure of the influence matrices is derived.

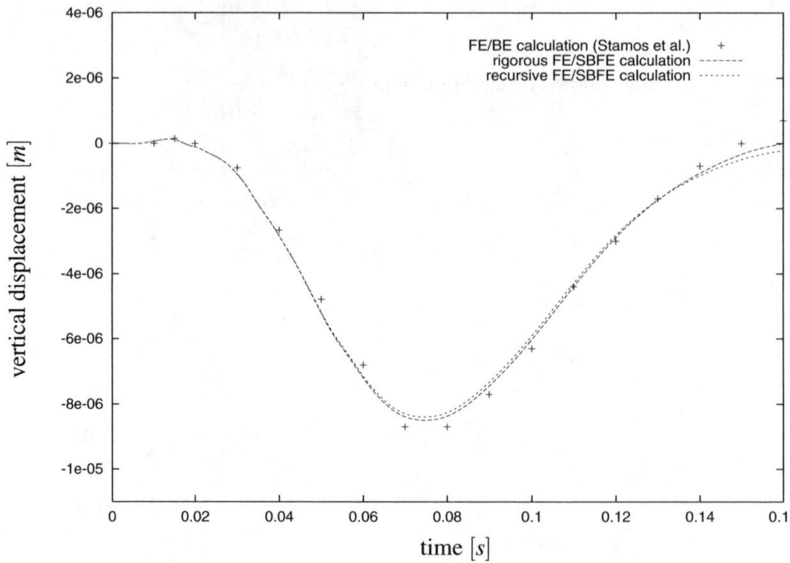

Fig. 4.16. Comparison of FE/BE and FE/SBFE procedures (coarse mesh, rigorous and recursive algorithm): computed vertical displacements at observation point

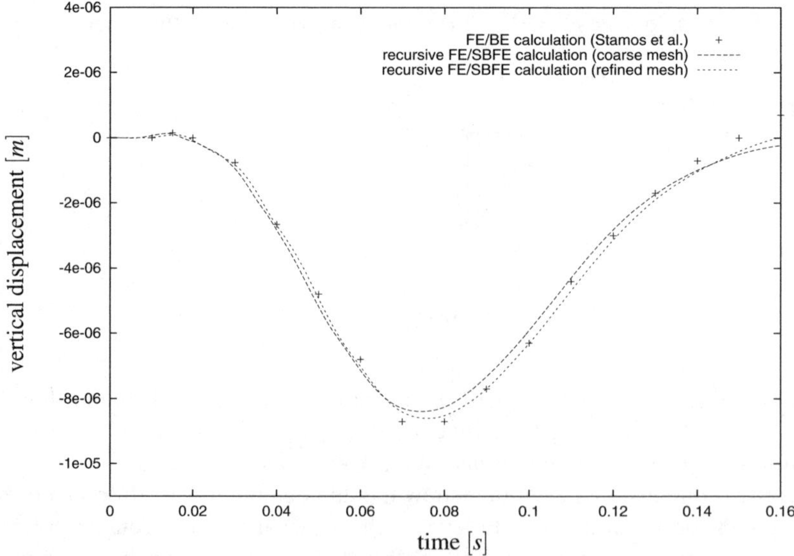

Fig. 4.17. Comparison of FE/BE and FE/SBFE procedures (recursive algorithm, coarse and refined mesh): computed vertical displacements at observation point

4.2 Three-dimensional benchmark example of soil-structure interaction in the time domain

In this subsection, a simple three-dimensional example illustrates the efficiency and accuracy of the coupled FE/SBFE method with the presented modifications. To compare the results, a BE approach and a coupled FE/SBFE approach with a coarse and fine mesh are analysed for two different load histories. The approximation in time and space of the SBFE approach is compared to calculations without approximation.

System

A concrete foundation with a Young's modulus $E = 30000MPa$, Poisson's ration $v = 0.20$, density $\rho = 2500\frac{kg}{m^3}$, length $5m$, width $3m$, and height $2m$, see Fig. 4.18, is horizontally loaded, and embedded in elastic soil. The soil has a Young's modulus of E=30MPa, Poisson's ratio $v = 0.30$, and a density $\rho = 1800\frac{kg}{m^3}$. Two different horizontal loads are applied to compare the computational effort of different numerical simulations: for load case I, a Heaviside-step load of intensity $4 \cdot 1.25MN = 5MN$ acts for $0.02s$ with a total simulation time of 0.1s, and secondly, a constant load of the same amount and a total simulation time of $5.0s$, is applied for load case II, see Fig. 4.19.

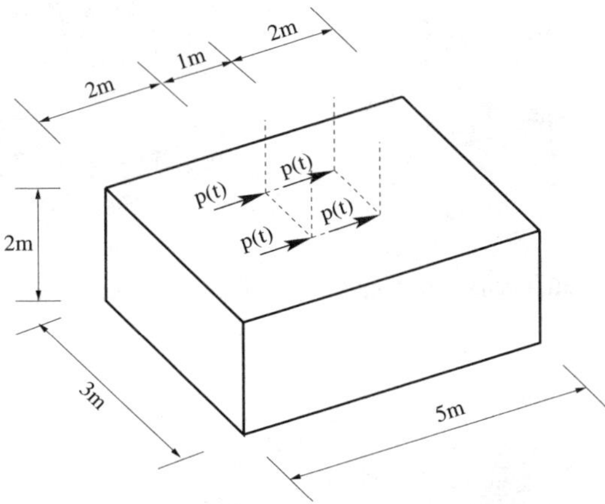

Fig. 4.18. Concrete foundation embedded in an elastic half-space under horizontal loading

Discretisation

The two applied FE discretisations for modelling the foundation are shown in Fig. 4.20. The coarse mesh consists of 30 eight-node hexagonal elements ($1m \times 1m \times 1m$) with a total of 216 degrees of freedom (DOFs), while the fine mesh is

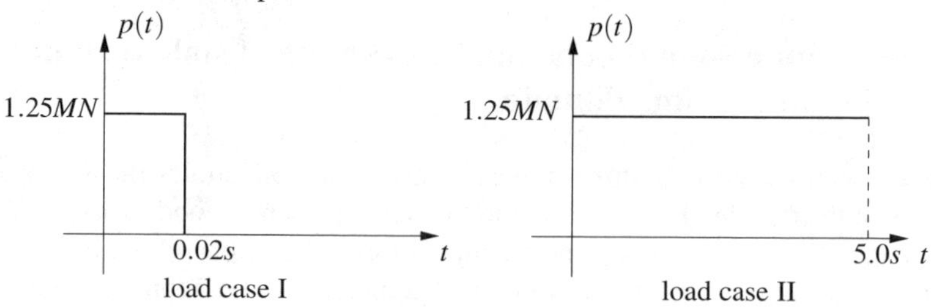

Fig. 4.19. Time-dependency of the horizontal load on foundation (load cases I and II)

built of 240 eight-node hexagonal elements (0.5m x 0.5m x 0.5m) with a total of 1155 DOFs. The SBFE part of the numerical model has 47 four-node plane elements with a total of 168 DOFs for the coarse mesh and 188 four-node plane elements with 615 DOFs for the refined mesh. Fig. 4.20 also shows the external forces, which are realised with four nodal loads in horizontal direction.

The pure boundary element approach consists of 124 three-node linear elements for the foundation with a total of 68 nodes and 334 elements for the elastic half-space surface (192 nodes), see Fig. 4.21.[3]

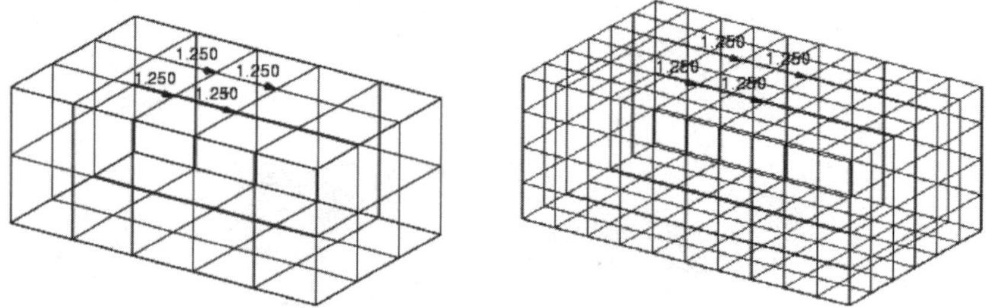

Fig. 4.20. Coarse and fine finite element mesh of foundation

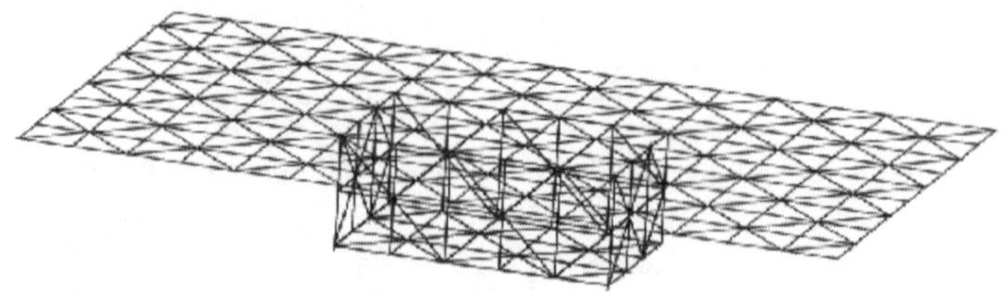

Fig. 4.21. Boundary element mesh

---

[3] This discretisation excludes infinite boundary elements.

Results

For load case I, the horizontal displacement of a loaded node is displayed in Fig. 4.22. These results, gained with a coupled FE/SBFE approach, a coarse mesh discretisation (see Fig. 4.20, left side) and varying time-step size, are compared to the boundary element results (a snapshot of computed displacements by the boundary element method is shown in Fig. 4.25). For the boundary element analysis, the convolution quadrature calculus of Lubich is used [93, 94, 116]. The calculated horizontal displacements are close to those of the boundary element approach, but there are some differences to the boundary element simulation for $t \geq 0.3s$. The chosen time-step size has no significant influence on the results.

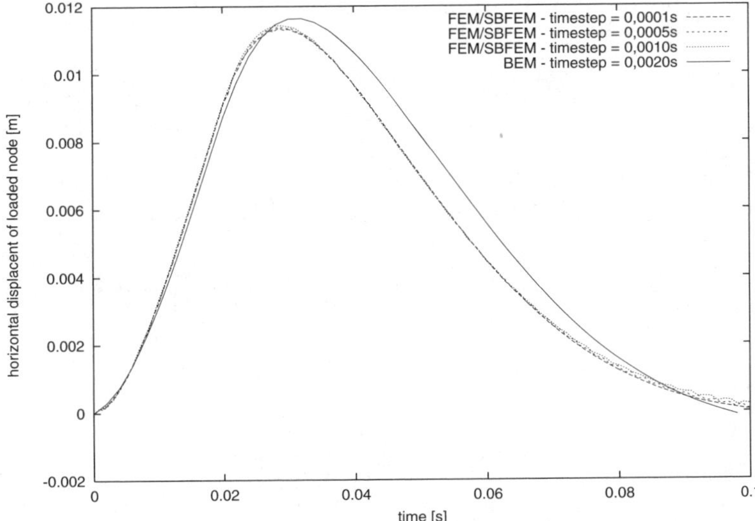

Fig. 4.22. Calculated horizontal displacement of a loaded node (load case I), BEM compared with coupled FE/SBFE approach (coarse mesh): influence of time-step size

A comparison of the results between a coarse and a fine mesh for the coupled FE/SBFE approach shows Fig. 4.23. The coupled FE/SBFE simulation with a refined mesh leads to results, which are a little bit closer to the BE solution.

Fig. 4.24 depicts results of load case I for a direct calculation of the convolution integral (3.85) and results gained by a calculation which is performed with the recursive algorithm of subsection 3.7. As results are identical, this demonstrates that it is possible to use the recursive algorithm presented in this work, without limitations and, moreover, no additional numerical parameters has to be chosen which is favourable.

Load case II is mainly chosen in order to show the necessary computational time for different analysis types. Tab. 4.1 displays the CPU time for the presented example

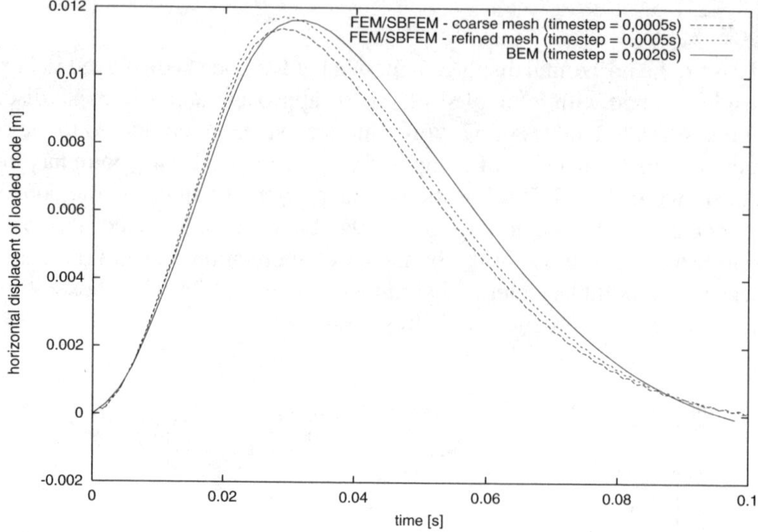

Fig. 4.23. Calculated horizontal displacement of a loaded node (load case I), BEM approach compared with coupled FE/SBFE approach: influence of the mesh refinement

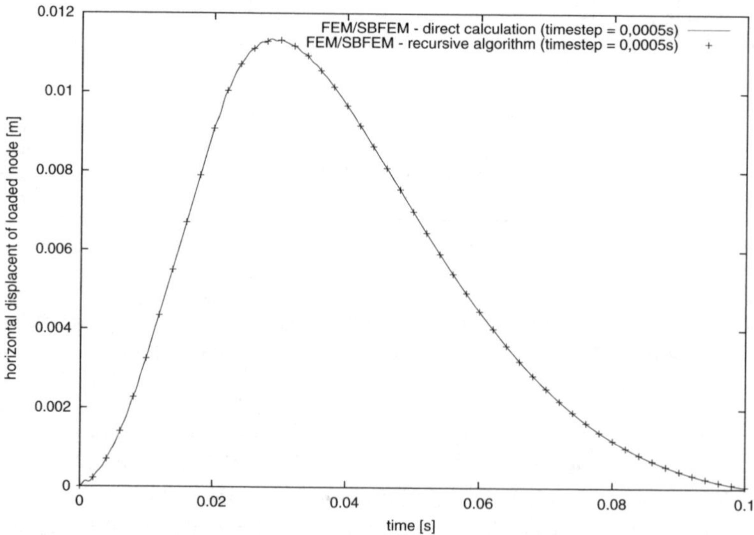

Fig. 4.24. Calculated horizontal displacement of a loaded node (load case I), FE/SBFE approach: direct calculation of convolution integral compared to the recursive algorithm

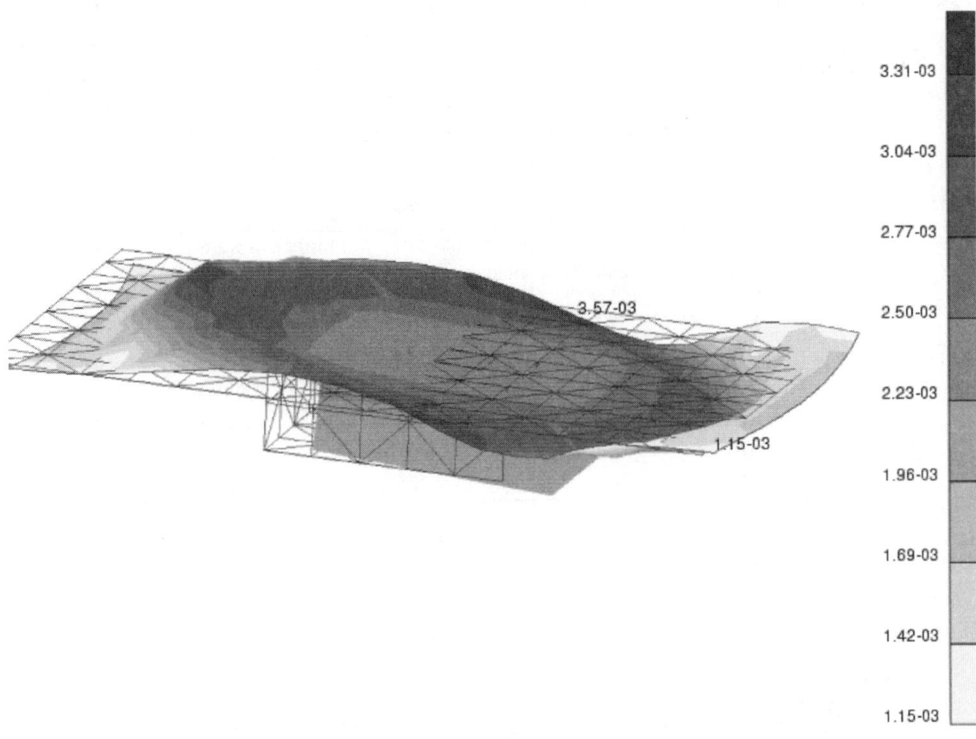

Fig. 4.25. Magnitude of displacements [m] for $t = 30 \cdot 0.0020s = 0.06s$ by the BEM

with coarse discretisation, load case II. All calculations have been performed on an UltraSparc II processor with a speed of $950MHz$. It is obvious that the rigorous calculation of the convolution integral is only applicable for short simulation times.

number of time-steps	CPU time FEM/SBFEM direct	CPU time FEM/SBFEM recursive
$[-]$	$[s]$	$[s]$
100	330	330
200	776	785
1000	32800	12500
5000	821000	52600
10000	3282000	127000

Table 4.1. CPU time for direct calculation and recursive algorithm (fine mesh)

For both discretisations Figs. 4.26 and 4.27 show the matrix population (where matrix entries, which are not equal to zero are marked with a dot) of the SBFE matrix for different zero-element thresholds. The banded structure of the matrix is evident for thresholds $\varepsilon_z \geq 10^{-4}$, especially for the refined mesh. The resulting matrix densities δ (defined as ratio of non-zero matrix entries to the total number of matrix entries) are collected in Tab. 4.2.

zero-element threshold [−]	type of mesh	total number of matrix entries [−]	number of non-zero elements [−]	density δ [%]
10^{-5}	coarse	28224	9644	34,2
10^{-4}	coarse	28224	5404	19,1
10^{-3}	coarse	28224	2180	7,7
10^{-5}	fine	378225	50813	13,4
10^{-4}	fine	378225	28647	7,6
10^{-3}	fine	378225	13507	3,6

Table 4.2. Matrix density δ for different zero-element thresholds

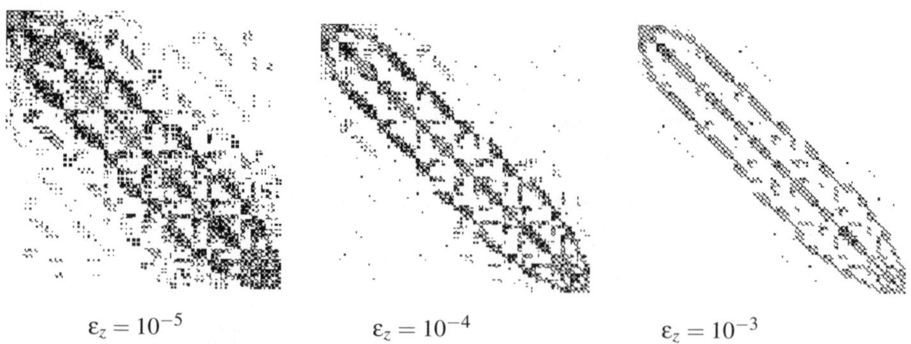

Fig. 4.26. SBFE influence matrix population of the coarse discretisation ($n = 28224$); different zero-element thresholds ε_z

For an accurate approximation with a zero-element threshold $\varepsilon_z = 10^{-4}$, the memory reduction for the coarse mesh is 80.9% and 90.4% for the refined mesh. Additionally, the CPU time consumption is reduced, because the matrix vector multiplication is less costly for a banded influence matrix.

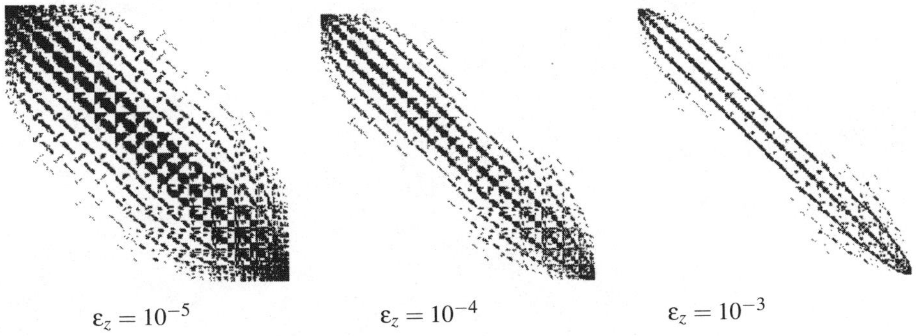

Fig. 4.27. SBFE influence matrix population of the refined discretisation ($n = 378225$); different zero-element thresholds ε_z

4.3 Concluding remarks

The computational effort can be significantly reduced with marginal loss of accuracy by using the recursive algorithm . This algorithm is based on an approximation of the unit acceleration response matrices by a linear time dependence after time t_m. This approximation results in an efficient scheme for the evaluation of the convolution integral. The results of the presented coupled FE/SBFE approach are close to the BEM for all simulations.

Part II

Applications

5. Wave propagation in fluids

In this section, a coupled FE/SBFE technique for the scalar wave equation and its application to wave propagation problems in fluids is presented.

A popular and efficient tool to handle dynamic structural problems is the FEM which leads to a symmetric system of equations to determine the unknown degrees of freedom (see Sec. 1). This domain type method is also often used to predict sound fields and it is especially suitable for finite and enclosed fluid domains [32].

When using discrete methods such as the FEM for the solution of so-called *Helmholtz problems* , one soon is confronted with the significance of the wave number k. The wave number characterises the oscillatory behaviour of the exact solution. The larger the value of k, the stronger the oscillations. To resolve this characteristic by the numerical model, a certain number of mesh points per wavelength has to be applied.[1] For large wavenumbers, the numerical solution is disturbed due to loss of operator stability at large wave numbers [78]. The examples in this section are limited to small wave numbers.

To determine wave propagation in *infinite domains*, several methods have been developed. These methods try to prevent energy from being reflected at the artificial boundary – namely the near-field/far-field interface – back into the domain. For example, the use of Dirichlet-to-Neumann boundary conditions [58] or of infinite elements [20, 43] fulfils this term, called Sommerfeld radiation condition . Other important families of these methods are the absorbing boundary conditions [69] and the perfectly matched layers [19, 1, 133, 72]. A detailed review on these methods is given by Givoli [59, 60], Tsynkov [132] and Thompson [131]. Another tool for simulating exterior problems is the well known BEM [7, 50], which fulfils the radiation condition implicitly. It has been used successfully for both standard and very specialised wave propagation problems, e.g. the sound radiation of moving sound sources like cars or trains [10] or outdoor sound problems [8, 70]. The use of the symmetric Galerkin boundary element method or variational boundary element method avoids the problem of non-symmetric system matrices which collocation produces, see, e.g., [26, 27, 81]. Furthermore, fast multipole boundary element methods were developed for getting solutions without exceeding high computational costs [39, 62, 104].

The numerical simulation of sound transmission through a component that separates an interior from an exterior domain is very important for practical applications [88]. For the coupling of the elastic structure and the interior fluid, a FE/FE coupling

---

[1] A minimal value of five mesh points per wavelength is widely accepted.

strategy will be applied. Furthermore, a FE/BE coupling methodology of the elastic structure with the exterior fluid domain is presented. An obvious approach to combine finite elements with boundary elements is an uncoupled one [83], i.e., the FE solution of a structural vibration analysis becomes the loading on the unbounded domain. Unfortunately by using such a method the real interaction phenomena between structural and fluid domain are lost. A more appropriate coupling starts from the principle of virtual work. Langer has applied this principle to the coupling of finite and boundary elements in the study of sound transmission through windows [88]. Alternatively, fluid-structure interaction phenomena can be considered using an iterative coupling procedure. Iterative mortar coupling schemes enable the use of different discretisations of different domains [53].

In the following, some results gained by a coupled FE/SBFE approach are compared to those of simulations based on the boundary element method for verification. Additionally, the previously presented methods for reducing the non-locality in time (Sec. 3.7.1) and space (Sec. 3.7.2), and the hierarchical matrices (Sec. 3.8) are applied.

5.1 Governing equations

In this subsection, some basic equations for the sound propagation in fluids, vibration of thin plates, and the FEM as well as the SBFEM associated with the scalar wave equation are given.

5.1.1 Sound propagation in unbounded fluids

Sound propagates in fluids as a pressure wave whose distribution satisfies the *scalar wave equation*

$$\Delta p(\mathbf{x},t) - \frac{1}{c^2}\ddot{p}(\mathbf{x},t) = 0 \tag{5.1}$$

in the time domain, where Δ is the Laplace operator, $p(\mathbf{x},t)$ is the time-dependent sound pressure, $\ddot{p}$ the acceleration of sound pressure p and c the speed of sound. If a time harmonic pressure $p(\mathbf{x},t) = \hat{p}(\mathbf{x})e^{i\omega t}$ is assumed, equation (5.1) can be transferred to the frequency domain, resulting in the well-known *Helmholtz equation* :

$$\Delta p + k^2 p = 0\,, \tag{5.2}$$

where $k = \frac{\omega}{c}$ is defined as wave number, i.e., the ratio of the circular frequency ω and the sound velocity c.

Additionally, for radiation problems, the *Sommerfeld condition* at infinity has to be fulfilled

$$\frac{\partial p}{\partial r} + c\dot{p} = O\left(\frac{1}{r}\right)\,, \quad r \to \infty\,, \tag{5.3}$$

for the time domain representation, or

$$\frac{\partial p}{\partial r} - ikp = O\left(\frac{1}{r}\right), \quad r \to \infty. \tag{5.4}$$

for the frequency domain.

5.1.2 Vibration of thin plates

For instance airborne sound waves induce bending vibrations of the separating components. These vibrations can be described by an appropriate plate theory depending on the slenderness of the separating component. Here, the *Kirchhoff differential equation* for plane plates [40] is used that is valid for thin components. So the bending vibrations can be described adequately by

$$B\Delta\Delta\eta + \rho h\omega^2\eta = p_B, \tag{5.5}$$

where η is the deflection , ρ the material density , h the thickness of the plate, p_B the plate loadings and B indicates the bending stiffness of the plate (5.13) with the material parameters Young's modulus E and the Poisson's ratio ν. For thicker separating components other plate theories have to be chosen, e.g., the Mindlin theory for moderately thick separating components [100, 37].

5.1.3 Finite element method

Vibrations can be treated numerically using the FEM, see Sec. 1. Here, the basic equations for frequency and time domain analysis of wave propagation in fluids, where the scalar wave equation applies, are given.

Frequency domain

By means of weighted residuals one obtains after discretisation the following governing equation for a fluid which is periodically stimulated with excitation frequency ω

$$\left(\mathbf{K}_f - k^2\mathbf{M}_f\right)\mathbf{p} = -\rho\omega^2\mathbf{d}, \tag{5.6}$$

where $\mathbf{K}_f$ and $\mathbf{M}_f$ are the compressibility matrix and the mass matrix of the fluid, respectively. The vector $\mathbf{d}$ represents the influence of sound sources and $\mathbf{p}$ indicates the nodal values of the sound pressure . For each finite element the sound pressure p can be described by $p = \mathbf{N}_a\mathbf{p}$ where $\mathbf{N}_a$ denotes the shape functions of the fluid domain. The compressibility matrix $\mathbf{K}_f^e$ of a single element of the fluid domain is computed using

$$\mathbf{K}_f^e = \int_V \nabla\mathbf{N}_a^T\nabla\mathbf{N}_a\,dV, \tag{5.7}$$

and the mass matrix $\mathbf{M}_f^e$ using

$$\mathbf{M}_f^e = \frac{1}{c^2}\int_V \mathbf{N}_a^T\mathbf{N}_a\,dV. \tag{5.8}$$

The FE formulation (5.6) is an effective method for treating interior problems as described above.

Again using weighted residues the *structural vibrations of plates* are modelled by the following FE equation

$$\left(\mathbf{K}_s - \omega^2 \mathbf{M}_s\right)\mathbf{u} = \mathbf{f} . \tag{5.9}$$

$\mathbf{K}_s$ and $\mathbf{M}_s$ denote the stiffness matrix and mass matrix of the structure, respectively, $\mathbf{f}$ is the load vector caused by plate loadings, $\mathbf{u}$ indicates the nodal values of the plate deflection η, approximated by $\eta = \mathbf{N}_s\mathbf{u}$ in each finite element of the discretised plate. The mass matrix $\mathbf{M}_s^e$ of each finite element is calculated using

$$\mathbf{M}_s^e = \rho_s h \int_A \mathbf{N}_s^T \mathbf{N}_s \, dA , \tag{5.10}$$

where ρ_s denotes the mass density of the plate and $\mathbf{N}_s$ are the shape functions of the structure. The element stiffness matrix $\mathbf{K}_s^e$ is computed as

$$\mathbf{K}_s^e = \int_A \mathbf{H}^T \mathbf{E} \mathbf{H} \, dA , \tag{5.11}$$

where $\mathbf{E}$ denotes the constitutive matrix for elastic material, with

$$\mathbf{E} = B \begin{bmatrix} 1 & v & 0 \\ v & 1 & 0 \\ 0 & 0 & \frac{1-v}{2} \end{bmatrix} , \tag{5.12}$$

where B is the bending stiffness of the plate, with

$$B = \frac{E}{1 - v^2} \frac{h^3}{12} , \tag{5.13}$$

$\mathbf{H}$ contains the derived shape functions

$$\mathbf{H} = \mathbf{D}\mathbf{N}_s . \tag{5.14}$$

$\mathbf{D}$ is the differential operator matrix

$$\mathbf{D}^T = \left[-\frac{\partial^2}{\partial x_1^2} \quad -\frac{\partial^2}{\partial x_2^2} \quad -\frac{\partial^2}{\partial x_1 x_2} \right] \tag{5.15}$$

that is applied to the shape functions $\mathbf{N}_s$ of the structure.

Time domain

The following governing equation in the time domain for a fluid which is stimulated with time-dependent flux $q(t)$ is obtained after discretisation, see,i.e., [50]:

$$\mathbf{K}\mathbf{p}(t) + \mathbf{M}\ddot{\mathbf{p}}(t) = \mathbf{q}(t) , \tag{5.16}$$

where $\mathbf{p}(t), \ddot{\mathbf{p}}(t)$ and $\mathbf{q}(t)$ indicate the vector of sound pressure, acceleration and flux, respectively. $\mathbf{K}$ is the compressibility matrix and $\mathbf{M}$ the mass matrix. The vector $\mathbf{q}$ represents the influence of sound sources and $\mathbf{p}$ indicates the nodal values of the sound pressure. For each finite element the sound pressure p can be described by $p = \mathbf{N}_a \mathbf{p}$, analogous to the frequency domain derivation. After assembly of all element contributions, (5.16) is obtained. It represents the equation of motion in the time domain for a general case.

If the domain is divided in a near-field and far-field part for coupled indoor (near-field) / outdoor (far-field) problems, and a time-dependent interaction flux vector $\mathbf{q}_r(t)$ (which represents the influence of the infinite domain) on the near-field/far-field interface is introduced, this equation can be rewritten as (equivalent to (3.128)):

$$\begin{bmatrix} \mathbf{K}_{ff} & \mathbf{K}_{fb} \\ \mathbf{K}_{bf} & \mathbf{K}_{bb} \end{bmatrix} \begin{bmatrix} \mathbf{p}_f(t) \\ \mathbf{p}_b(t) \end{bmatrix} + \begin{bmatrix} \mathbf{M}_{ff} & \mathbf{M}_{fb} \\ \mathbf{M}_{bf} & \mathbf{M}_{bb} + \gamma \Delta t \mathbf{M}_0^\infty \end{bmatrix} \begin{bmatrix} \ddot{\mathbf{p}}_f(t) \\ \ddot{\mathbf{p}}_b(t) \end{bmatrix} = \begin{bmatrix} \mathbf{q}_f(t) \\ \mathbf{q}_b(t) - \mathbf{q}_r(t) \end{bmatrix},$$
$$(5.17)$$

where the compressibility matrix $\mathbf{K}$, the mass matrix $\mathbf{M}$, and the nodal value vectors of the pressures $\mathbf{p}$ and accelerations $\ddot{\mathbf{p}}$, respectively, are subdivided corresponding to the location of the nodes, i.e., the subscript b denotes the nodes on the near-field/far-field interface (boundary) and the subscript f the remaining nodes of the fluid.

The vector $\mathbf{q}_r(t)$ can be determined by the convolution integral (similar to the interaction force-acceleration relationship of Sec. 3.5, equation (3.85)):

$$\mathbf{q}_r(t) = \int_0^t \mathbf{M}^\infty(t - \tau)\ddot{\mathbf{p}}(\tau)\, d\tau\,, \qquad (5.18)$$

with $\mathbf{M}^\infty$ being the acceleration unit-impulse response matrix , which can be obtained by the SBFEM in the time domain corresponding to the acceleration unit-impulse response matrix of the vector wave equation, see Wolf and Song [147].

5.1.4 Scaled boundary finite element method

Following the same concept as for the vector wave equation , within the transformation of the governing scalar wave equation of acoustics (5.1) in the scaled boundary coordinate system, only the effect of the Laplace operator needs to be considered.

5.2 Implementation

In the following, some implementation details of the SBFEM in the frequency domain and time domain are described. Furthermore, some special aspects of the required coupling procedures are explained.

5.2.1 Scaled boundary finite element method in the frequency domain

The SBFE equation formulated in the *frequency domain* is derived using the relationship of the dynamic stiffness matrices based on similarity formulated in the continuous form and the relationship based on assemblage of finite element cell and unbounded medium [147].

To couple the SBFE and FE domains, only the previous calculated frequency dependent symmetric SBFE matrices $\mathbf{S}^{\infty}(\omega_i)$ need to be added to the system matrix of the FE domain. The following solution procedure is the same as for a pure FEM calculation, therefore the scaled boundary approach can easily be incorporated in an existing FE computer code.

The standard FE equation (5.6) is extended as a function of a discrete frequency ω_i:

$$(\mathbf{K} - k^2\mathbf{M} + \mathbf{S}^{\infty}(\omega_i))\mathbf{p} = -\rho\omega_i^2\mathbf{d} . \tag{5.19}$$

The addition of $\mathbf{S}^{\infty}(\omega_i)$ to the system matrix does not change the standard FE solution procedure. Equation (5.19) represents the FE/SBFE coupling of the fluid.

5.2.2 Scaled boundary finite element method in the time domain

For a numerical treatment in the *time domain* , a time discretisation of the convolution integral (5.18) is needed. If a piecewise constant approximation of the acceleration unit-impulse response matrix is assumed, see (3.124) of Sec. 3.5.1, the discrete form of (5.18) is obtained:

$$\mathbf{q}_r(t_i) = \sum_{j=1}^{i} \mathbf{M}_{i-j}^{\infty} \int_{(j-1)\Delta t}^{j\Delta t} \ddot{\mathbf{p}}(\tau)\,d\tau = \sum_{j=1}^{i} \mathbf{M}_{i-j}^{\infty}(\dot{\mathbf{p}}(t_j) - \dot{\mathbf{p}}(t_{j-1})) . \tag{5.20}$$

Introducing the γ-parameter of the Hilber-Hughes-Taylor implicit time integration scheme [75], which is used to solve (5.17), and separating the unknown acceleration vector $\ddot{\mathbf{p}}(t_i)$, the calculation of $\mathbf{q}_r(t_i)$ can be performed with the following equation:

$$\mathbf{q}_r(t_i) = \gamma\Delta t\mathbf{M}_0^{\infty}\ddot{\mathbf{p}}(t_i) + \sum_{j=1}^{i-1} \mathbf{M}_{i-j}^{\infty}(\dot{\mathbf{p}}(t_j) - \dot{\mathbf{p}}(t_{j-1})) . \tag{5.21}$$

Thus, (5.17) can be rewritten as:

$$\begin{bmatrix} \mathbf{K}_{ff} & \mathbf{K}_{fb} \\ \mathbf{K}_{bf} & \mathbf{K}_{bb} \end{bmatrix} \begin{bmatrix} \mathbf{p}_f(t) \\ \mathbf{p}_b(t) \end{bmatrix} + \begin{bmatrix} \mathbf{M}_{ff} & \mathbf{M}_{fb} \\ \mathbf{M}_{bf} & \mathbf{M}_{bb} + \gamma\Delta t\mathbf{M}_0^{\infty} \end{bmatrix} \begin{bmatrix} \ddot{\mathbf{p}}_f(t_i) \\ \ddot{\mathbf{p}}_b(t_i) \end{bmatrix}$$
$$= \begin{bmatrix} \mathbf{q}_f(t_i) \\ \mathbf{q}_b(t_i) - \sum_{j=1}^{i-1} \mathbf{M}_{i-j}^{\infty}(\dot{\mathbf{p}}_r(t_j) - \dot{\mathbf{p}}_r(t_{j-1})) \end{bmatrix} . \tag{5.22}$$

For large i (i.e., a long simulation time) the direct solution of (5.21) is very time consuming, but an approximation in time leads to a fast recursive algorithm , see Sec. 3.7.1 or [91]. Another way to improve the efficiency of the SBFEM is to reduce the non-locality in space by omitting near-zero elements of the influence matrices $\mathbf{M}_i^{\infty}$ which results in sparse influence matrices, see Sec. 3.7.2 and [89]. An alternative way, the use of hierarchical matrices is described in Sec. 3.8 and in [92].

5.2.3 Boundary element method

The BEM is used in Sec. 5.4.2 as a reference. It is based on an integral representation of the governing equation. The boundary integral representation of the sound pressure is given by

$$c(\xi)p(\xi) + \int_\Gamma q^*(\mathbf{x},\xi;k)p(\mathbf{x})\,d\Gamma$$
$$= \int_\Gamma p^*(\mathbf{x},\xi;k)q(\mathbf{x})\,d\Gamma + \int_\Omega a(\mathbf{x})p^*(\mathbf{x},\xi;k)\,d\Omega\,, \qquad (5.23)$$

where p^* is the fundamental solution of the Helmholtz equation and q^* its singular flux. Both are implicitly dependent on the wave number k. a indicates the intensity of sound sources in the domain. The discretisation of the boundary Γ and approximation of the sound pressure and the sound flux with $p = \mathbf{N}_p\mathbf{p}$ and $q = \mathbf{N}_q\mathbf{q}$, respectively, on each separate element leads with help of the collocation at every boundary node to the standard *boundary element equation*

$$\mathbf{Gq} - \mathbf{Hp} = \mathbf{0}\,, \qquad (5.24)$$

where for abbreviation $a(\mathbf{x}) \equiv 0$ (no sound sources inside the domain) is assumed. The sound pressure $\mathbf{p}$ is determined at a Neumann boundary from the known (or given) sound flux $\bar{\mathbf{q}}$, while the sound flux $\mathbf{q}$ is determined at a Dirichlet boundary from the known sound pressure $\bar{\mathbf{p}}$.

5.3 Coupling of fluid and structure

In this subsection, the coupling of the SBFEM for wave propagation problems with the FEM for an elastic structure is presented. The achieved numerical algorithm is best suited to study the wave propagation in an unbounded domain or interaction phenomena of a vibrating structure and an unbounded fluid (acoustical) domain. For an acoustic analysis of an elastic structure that is embedded in or in contact with a fluid it is advantageous to use the FEM for structures and finite acoustic domains, and the SBFEM or BEM for exterior infinite domains.

5.3.1 FE/FE coupling of elastic structure and interior fluid

For an analysis of the interactions between a plate and a finite fluid domain, (5.6) and (5.9) have to be coupled. An alternative way to derive the FE equations is *Hamilton's principle* . This variation equation leads to a coupled system of equation

$$\delta(E_{kin} - E_{pot}) + \delta W = 0\,. \qquad (5.25)$$

Considering the variation of kinetic energy E_{kin} and potential energy E_{pot} yields the mass matrices and stiffness/compressibility matrices of the FE systems (5.6) and

(5.9), while the term δW of (5.25) leads to a coupling of acoustical and structural degrees of freedom at the interface. Here, the product of the sound pressure and of the plate deflection has to be integrated over the interface Γ_I (subscript I stands for interface) which means for the fluid finite element equation (5.6)

$$\delta W = -\int \delta p \, \eta \, d\Gamma_I \,, \tag{5.26}$$

since there the sound pressure is varied. Analogously, the virtual work at the interface with the structure finite element equation (5.9) is expressed by

$$\delta W = -\int p \, \delta \eta \, d\Gamma_I \,. \tag{5.27}$$

To implement the coupling, the deformation degrees of freedom $\mathbf{u} = [w, w_{,1}, w_{,2}, w_{,12}]$ of the structure have to be transformed so that only the deflection degree of freedom w remains. The corresponding vector of the nodal values is called $\mathbf{w}$. Finally, the following coupled system of equations for the fluid pressure (subscript f) and the plate deflection (subscript s – structure) is obtained:

$$\left(\mathbf{K}_f - k^2 \mathbf{M}_f\right) \mathbf{p}_f = -\rho \omega^2 \mathbf{C}_{FE/FE}^T \mathbf{w}_{sI} \tag{5.28}$$

$$\left(\mathbf{K}_s - \omega^2 \mathbf{M}_s\right)^* \mathbf{w}_s = -\mathbf{C}_{FE/FE} \mathbf{p}_I \,, \tag{5.29}$$

where (5.28) is the finite element description of the fluid pressure with the coupling matrix $\mathbf{C}_{FE/FE}$ and (5.29) the description of the structure deflection, with $\left(\mathbf{K}_s - \omega^2 \mathbf{M}_s\right)^*$ is the condensed dynamic stiffness matrix of the plate. $\mathbf{w}_I$ and $\mathbf{p}_I$ indicate the node values of the deflections and the sound pressure at the interface, respectively. If both sides of the plate are in contact with a fluid, $\mathbf{p}_I = \mathbf{p}_{If_1} - \mathbf{p}_{If_2}$ represents the pressure jump at the interface. $\mathbf{C}_{FE/FE} = \int \mathbf{N}_s^T \mathbf{N}_a \, d\Gamma_I$ is the coupling matrix of the interface. With (5.28) and (5.29) the coupling of FE modelled components is established.

5.3.2 FE/SBFE coupling of elastic structure and exterior fluid

Elastic structures that interact with unbounded fluid domains can be treated analogously to the above described coupled FE/FE system of equations (5.28) and (5.29) with a coupled FE/SBFE algorithm. The governing equation for the fluid pressure distribution (5.29) has to be replaced by the SBFE equation (5.19) to get the coupled FE/SBFE set of equations (5.30) and (5.31).

The adequate SBFE equation for the fluid reads

$$\left(\mathbf{K}_f - k^2 \mathbf{M}_f + \mathbf{S}^\infty(\omega_i)\right) \mathbf{p}_f = -\rho \omega^2 \mathbf{C}_{FE/FE}^T \mathbf{w}_{Is} \,, \tag{5.30}$$

and for the structure (plate):

$$\left(\mathbf{K}_s - \omega^2 \mathbf{M}_s\right)^* \mathbf{w}_s = -\mathbf{C}_{FE/FE} \mathbf{p}_I \,. \tag{5.31}$$

5.3.3 FE/BE coupling of elastic structure and exterior fluid

The FE/BE coupled solution of the sound transmission problem will be referred to as
a reference in Sec. 5.4.1. Therefore, the coupling between the plate and the boundary
element formulation of the exterior is given in the following.

The degrees of freedom of the different parts are the pressure of the FE handled
fluid domains, the displacements of the plates, and the sound pressure as well as the
sound flux in the boundary element treated domain.

The non-interface boundary must be of Dirichlet or Neumann type. The interface
merges the FE with the BE domain, so neither flux nor pressure is known a priori. At
the interface, accelerations of elastic and fluid particles have to be identical. Thus,
coupling of sound flux and deflection can be formulated as:

$$\mathbf{q} = \frac{\partial p}{\partial n} = \rho\omega^2\mathbf{w}.$$ (5.32)

Applying the principle of virtual work, one obtains for the coupling of the sound
pressure

$$\delta W = -\int p\,\delta\eta\,d\Gamma_I\,.$$ (5.33)

Finally, the following set of algebraic equations is obtained:

$$\left(\mathbf{K}_f - k^2\mathbf{M}_f\right)\mathbf{p}_f = -\rho\omega^2\mathbf{C}^T_{FE/BE}\mathbf{w}_{Is} \qquad \text{FE-fluid,}$$ (5.34)

$$\left(\mathbf{K}_s - \omega^2\mathbf{M}_s\right)^*\mathbf{w}_s = \mathbf{C}_{FE/BE}\mathbf{p}_I^\infty - \mathbf{C}_{FE/FE}\mathbf{p}_{If} \qquad \text{FE-plate,}$$ (5.35)

$$\rho_F\omega^2\mathbf{G}\mathbf{w}_s - \mathbf{H}\mathbf{p}^\infty = \mathbf{0} \qquad \text{BE-fluid – interface,}$$ (5.36)

$$\mathbf{G}\mathbf{q}^\infty - \mathbf{H}\mathbf{p}^\infty = \mathbf{0} \qquad \text{BE-fluid – non-interface.}$$ (5.37)

In the case of using at the interface the same shape functions for both, the finite and
the boundary elements, the coupling matrices $\mathbf{C}_{FE/BE}$ and $\mathbf{C}_{FE/FE}$ are equal.

5.4 Numerical examples

In the following, the SBFEM is applied to study the sound transmission through
a separating component, for the determination of the sound field around a sound
insulating wall, and, finally, for the sound propagation in a waveguide. The results are
compared to a hybrid algorithm of finite and boundary elements or to the boundary
element method, respectively.

In the first example a finite cuboid is coupled with the acoustic-medium space
through a glass plate, in the second example a waveguide is analysed, while in the
third example a pure exterior problem is modelled. The simulation of the sound field
originating from a sound source located behind a sound barrier is performed. This
system is analysed in the frequency and time domain.

5.4.1 Fluid-structure-fluid interaction (frequency domain analysis)

For the first example, an interior fluid-filled space is coupled through a glass plate (simulated using Kirchhoff plate elements) to the acoustic-medium space where the Sommerfeld radiation condition has to be fulfilled. The coupling is realised on the interface where the fluid and the structure parts are in contact (see Fig. 5.1).

System

The geometric definition of the first example is shown in Fig. 5.1. It consists of a cuboid filled with air (density $\rho = 1.21\frac{kg}{m^3}$, speed of sound $c = 346\frac{m}{s}$). The cuboid has acoustically rigid side walls. One face side is excited with a sound impact of $1\frac{m}{s}$ and the other side wall is coupled with a glass plate (Young's modulus $E = 6320MPa$, Poisson's ratio $\nu = 0.24$, density $\rho = 2300\frac{kg}{m^3}$ and thickness $h = 0.01m$). This glass plate is also coupled with an acoustic medium-space of air.

This example is chosen because results of a coupled FE/BE calculation with geometry shown in Fig. 5.1 can be found in Langer [87].

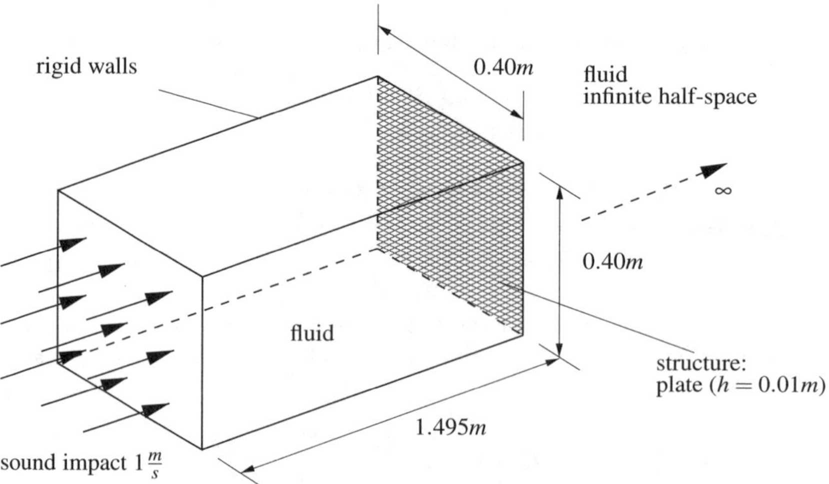

Fig. 5.1. Fluid-structure-fluid interaction: definition of problem

Discretisation

A schematic side-view of the discretisation for the coupled FE/SBFE simulation is shown in Fig. 5.2. The eight-node three-dimensional fluid elements of the interior fluid-filled space are coupled with four-node Kirchhoff elements. These plate elements are also coupled with a layer of auxiliary finite elements. This auxiliary layer is introduced to the system in order to carry the scaled boundary finite elements which represent the acoustic-medium space of the fluid. Thus, the coupling of FE/SBFE fluid can be simply established by equation (5.19).

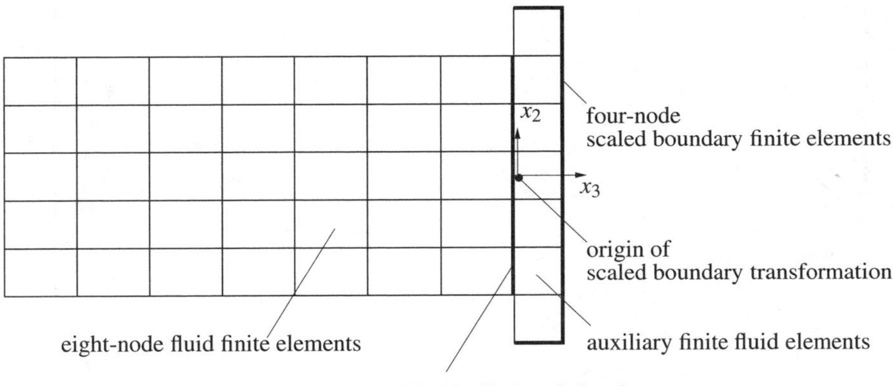

Fig. 5.2. Schematic side-view of discretisation (fluid-structure-fluid interaction problem)

The three-dimensional FE mesh of the fluid-structure-fluid interaction problem consists of $5 \times 5 \times 12 = 300$ three-dimensional eight-node fluid elements, $5 \times 5 = 25$ four-node Kirchhoff plate elements, $7 \times 7 = 49$ auxiliary eight-node fluid elements and 49 four-node two-dimensional scaled boundary finite elements.

Results

The computation is carried out in a frequency range of $90Hz$ to $200Hz$. Fig. 5.3 shows the calculated sound pressure level inside the cuboid (interior fluid-filled space) and in front of the glass plate (exterior space) as a function of the excitation frequency. To compare the described numerical method with a coupled FE/BE method, results of both methods are presented. The results show no major deviations, especially the eigenfrequencies are nearly identical.

5.4.2 Sound insulating wall (frequency and time domain analysis)

For the next example, the sound radiation of a simplified sound source (for example due to a car) behind a barrier (sound insulating wall) is simulated. All objects are considered to be acoustically rigid. So no interaction between the fluid (air) and the structure (sound barrier, building, and ground) is taken into account.

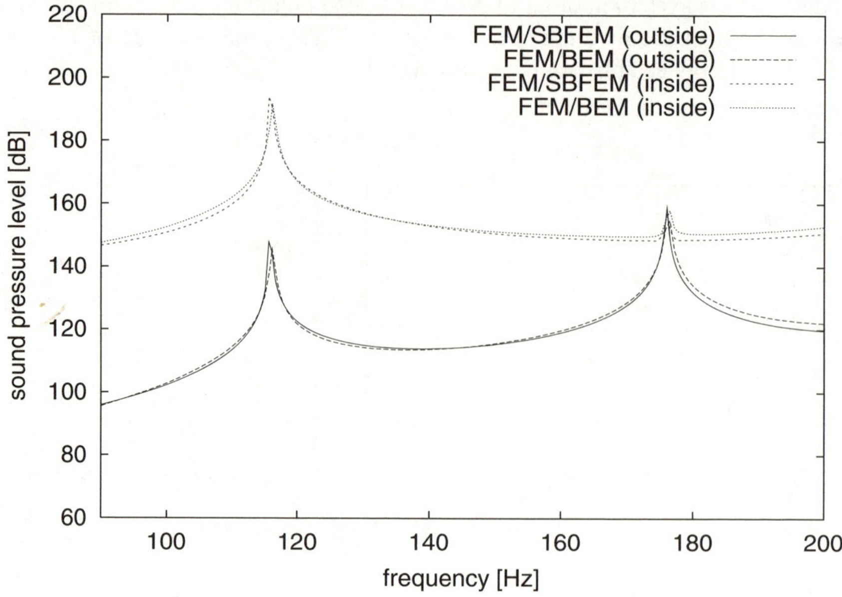

Fig. 5.3. Calculated sound pressure level in cuboid (inside) and in front of glass plate (outside) for coupled FE/BE and FE/SBFE procedure versus frequency

System

Fig. 5.4 shows the top view and cross section of the simulated domain. The sound source is located in a distance of $2m$ behind the sound insulating wall ($1m$ thick, $3m$ high and $30m$ long), while a building ($4 \times 5 \times 9m^3$) is located on the other side of the wall in $6m$ distance. The surrounding fluid is assumed to be air. The sound source is given as Neumann boundary condition ($v_n = 5\frac{m}{s}$) for the frequency domain analysis which represents a simplified assumption for vibrating side sheets of a car. For the time domain calculation, the sound source is given as sound pressure (Dirichlet boundary condition, see Fig. 5.5).

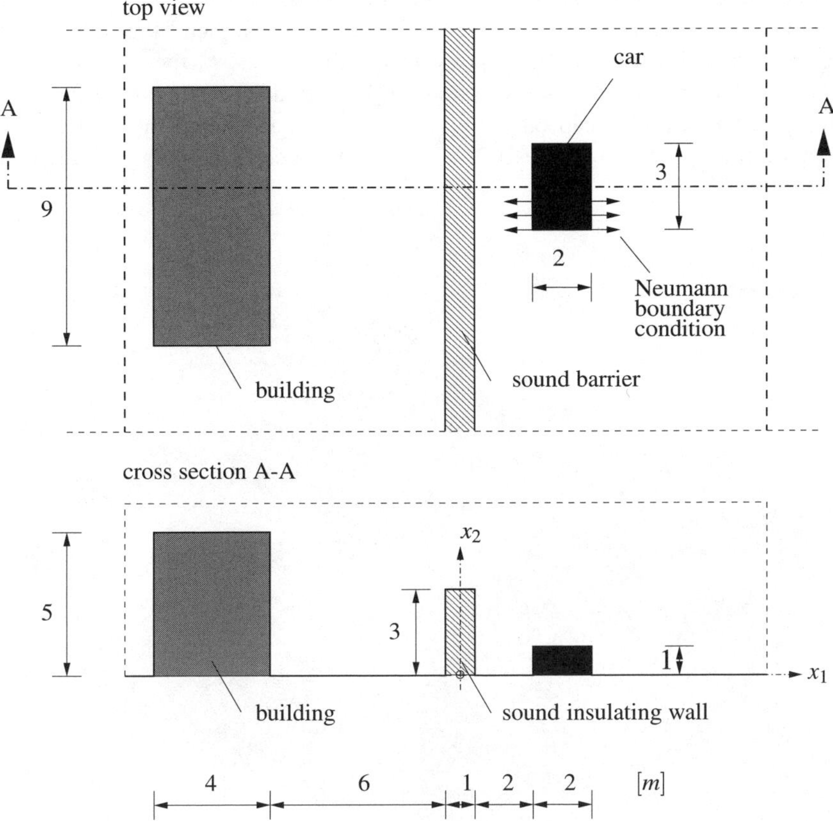

Fig. 5.4. Schematic top view and cross section of problem definition (exterior domain)

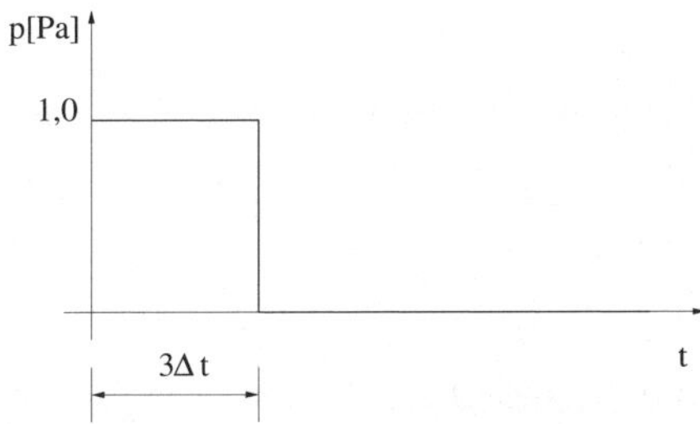

Fig. 5.5. Time-dependent pressure for time domain analysis

Discretisation

For the boundary element analysis and the coupled procedure, both element meshes are shown in Figs. 5.6 and 5.7, respectively. While the coupled FE/SBFE technique (Fig. 5.6) provides results at any point inside the FE mesh, the results of a pure boundary element calculation are found only at the discretised boundary. The solution at internal points can only be obtained on an additional mesh, where these internal points have to be placed in a certain distance to boundary elements (see Fig. 5.7).

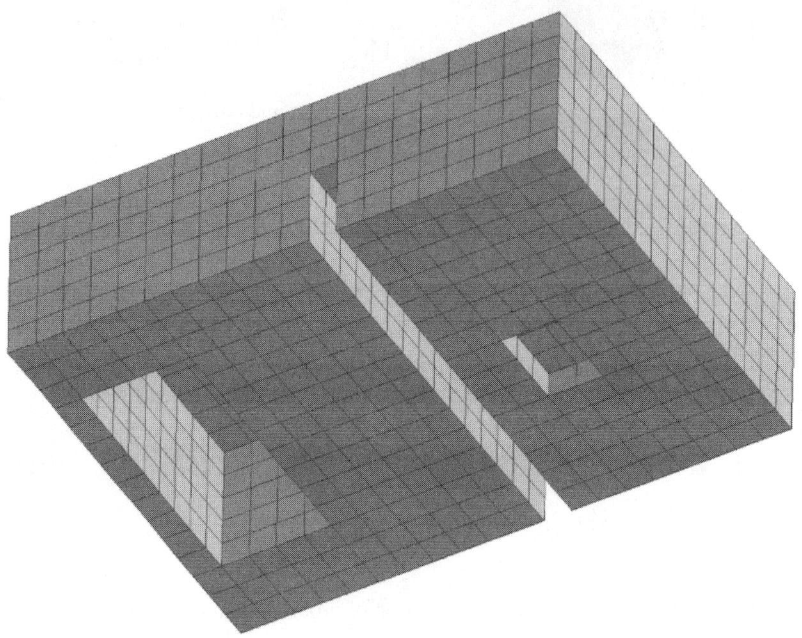

Fig. 5.6. FE mesh used for the coupled FE/SBFE simulation – bottom view

Frequency domain

The calculated logarithmic values of amplitudes of the sound pressure at cross section A-A for an excitation frequency of $20Hz$ are presented in Fig. 5.8. Differences in geometry of both displayed areas result from the fact that internal points from boundary element calculation must have a distance to boundary elements (see above).

Fig. 5.9 shows the calculated sound pressure at cross section A-A in height $y = 4m$ as a function of the x_1-coordinate (coordinate system see Fig. 5.4): an excellent agreement between both methods is observable.

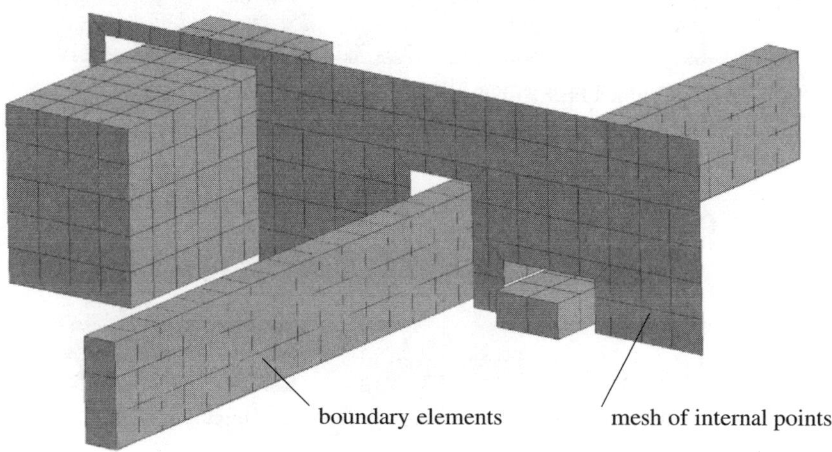

Fig. 5.7. BE mesh and mesh of internal points

The CPU time consumption for both calculations is similar. But for different load cases or geometries within the near-field (buildings, wall etc.), the stored frequency dependent matrices $S^\infty(\omega_i)$ of (5.19) stay the same. Thus, additional simulations can be carried out without effort for different near-field geometries or other load cases.

In Fig. 5.10 (coupled procedure) and Fig. 5.11 (BEM) a three-dimensional plot of the calculated results (logarithmic value of amplitudes of sound pressure at excitation frequency of $20Hz$) is presented. To visualise the sound pressure at cross-section A-A, the cutted FE mesh is displayed in Fig. 5.10). In both plots, the influence of the insulating wall is obvious.

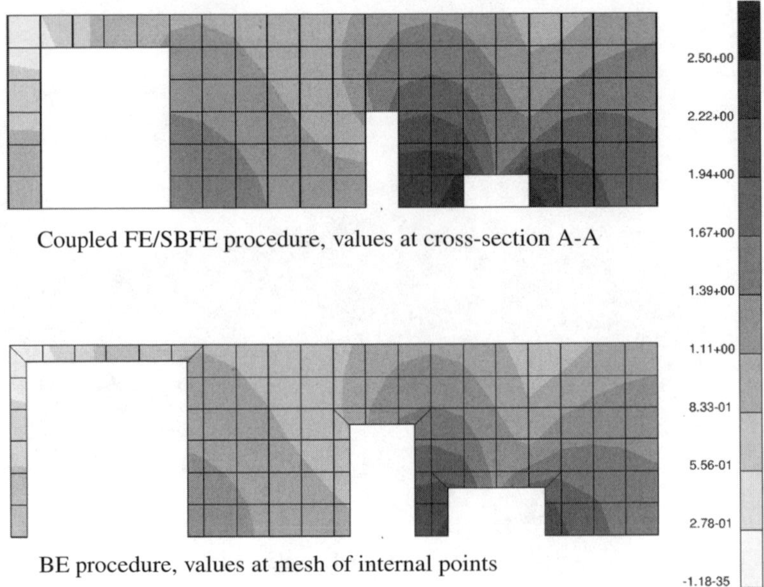

Fig. 5.8. Logarithmic value of sound pressure amplitude for excitation frequency of $20Hz$ calculated with coupled FE/SBFE and with BE procedure

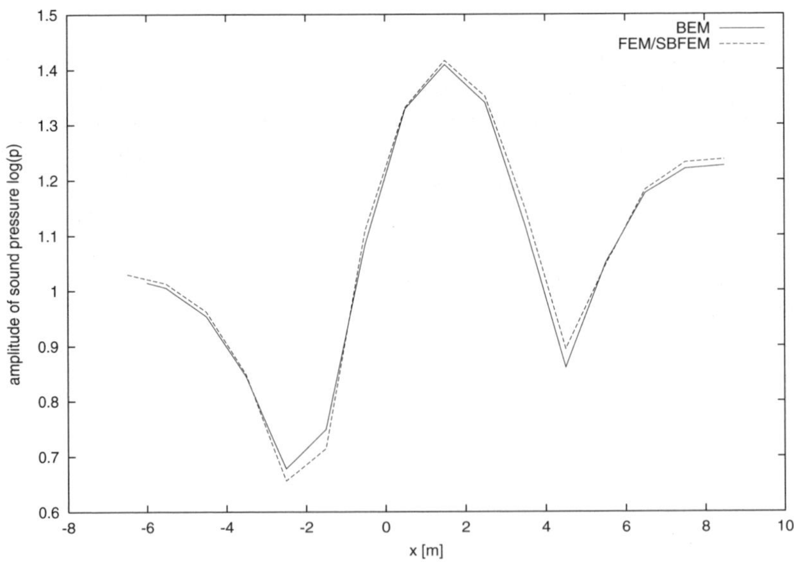

Fig. 5.9. Logarithmic value of sound pressure amplitude for excitation frequency of $20Hz$ calculated with coupled FE/SBFE and BE procedure, middle plane at $x_2 = 4m$

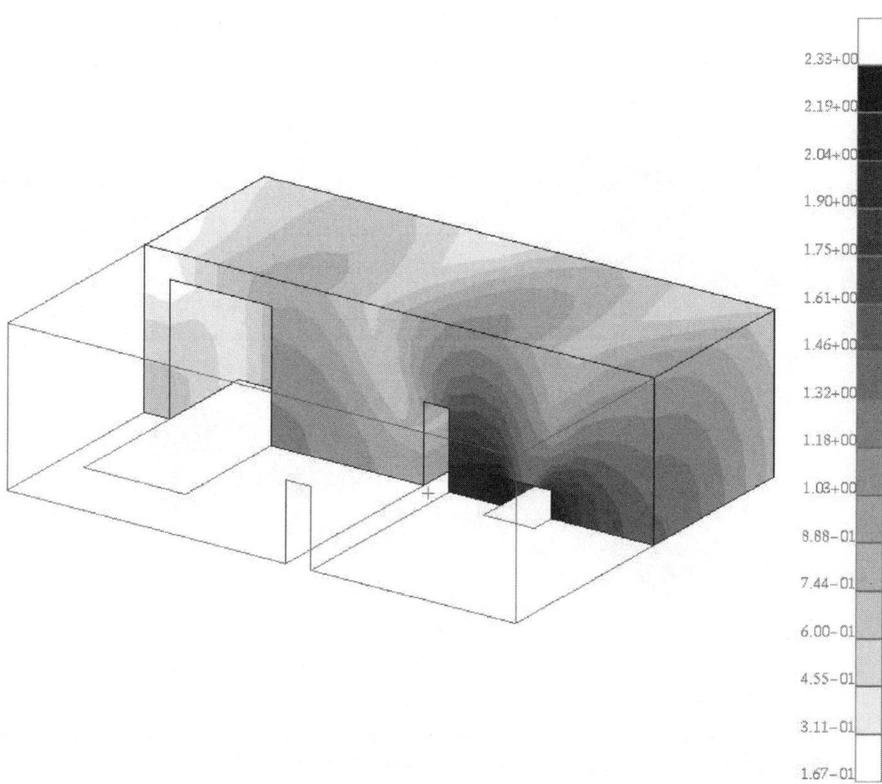

Fig. 5.10. Logarithmic value of sound pressure amplitude for excitation frequency of $20Hz$ calculated with coupled FE/SBFE procedure – three-dimensional view of one half of the mesh

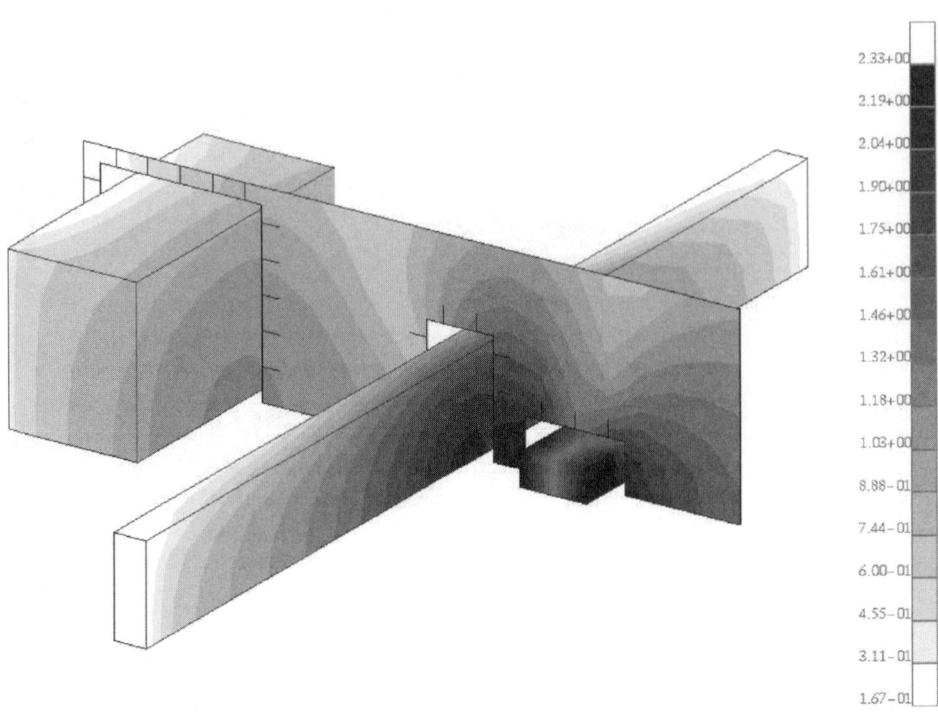

Fig. 5.11. Logarithmic value of sound pressure amplitude for excitation frequency of $20Hz$ calculated with BE procedure at boundary elements and internal points mesh

Time domain

For a finite element analysis a time-step $\Delta t \leq \frac{\min \ell_e}{10c}$ is suitable, where c is the wave speed and $\min \ell_e$ the smallest finite element length, which results in $\Delta t \leq \frac{1.0m}{10 \cdot 346 \frac{m}{s}} = 2.89 \cdot 10^{-4}s$; a constant time-step $\Delta t = 2 \cdot 10^{-4}s$ is chosen. The time-dependent pressure boundary condition is given in Fig. 5.5. Figs. 5.12 and 5.13 show the calculated pressure at a point in front of the sound insulating wall and in front of the building, respectively. The computation is carried out with a rigorous (direct) coupled FE/SBFE approach (see Sec. 3.6) and, alternatively, with application of the concept of hierarchical matrices . For both locations, the simulation with $\mathcal{H}$-matrix approximation generates the same results as the direct coupled FE/SBFE approach. The CPU time consumption is highly reduced: from $631s$ (evaluation of the convolution integral, 250 time steps, direct calculation) to $245s$ ($\mathcal{H}$-matrix approximation), see Fig. 5.14.

In Fig. 5.15 the structure of the $\mathcal{H}$-matrix for an influence matrix of the sound barrier example is shown. The dark blocks are inadmissible leafs (block matrices, which cannot be represented by a low rank approximation, see Sec. 3.8), while the light blocks are low rank representations of the corresponding blocks. The numbers in these blocks are the ranks of the block matrices ($\mathrm{rank}(\mathbf{A}_{ij}$ as introduced in Sec. 3.8). For this example, the maximum rank k of a matrix block is set to 20.

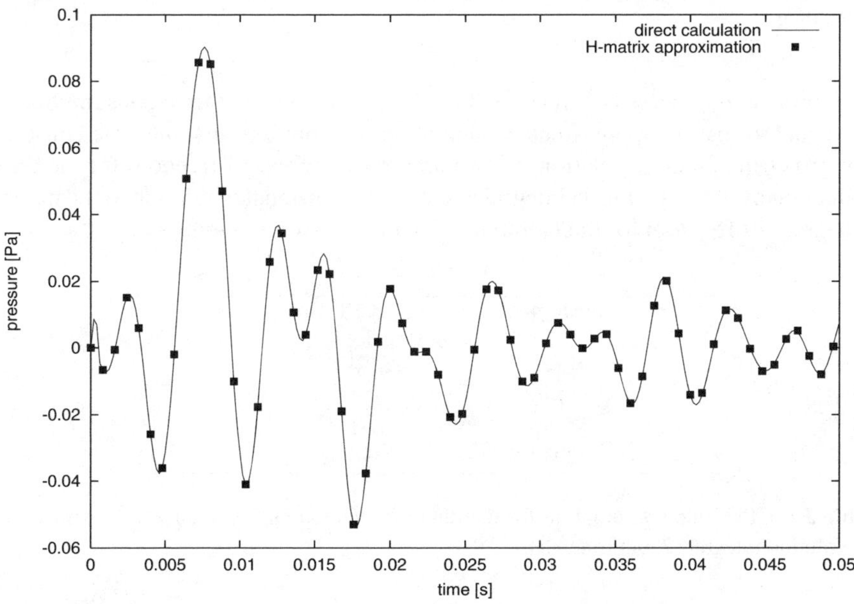

Fig. 5.12. Comparison of direct calculation and $\mathcal{H}$-matrix approximation for pressure in front of sound insulating wall

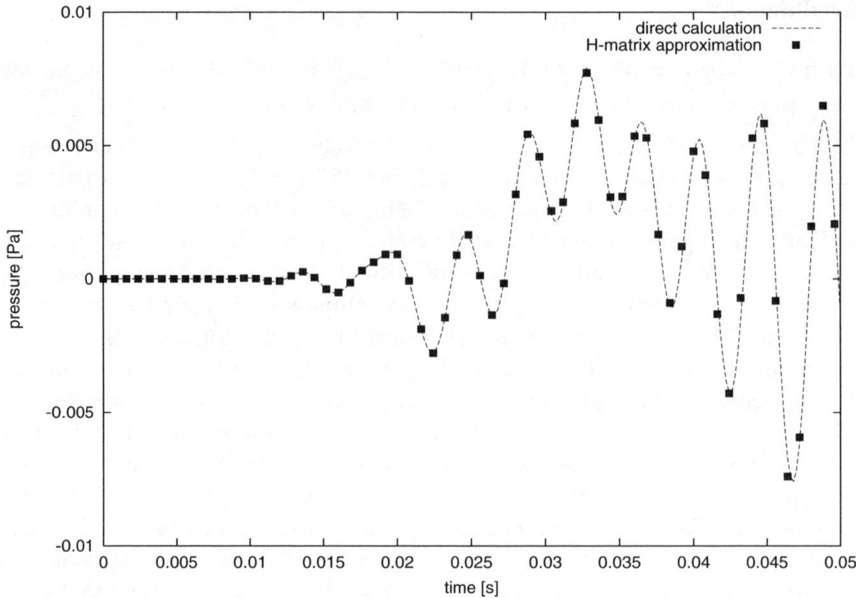

Fig. 5.13. Comparison of direct calculation and $\mathcal{H}$-matrix approximation for pressure in front of building

For different degrees of freedom Tab. 5.1 shows the CPU time consumption for direct and $\mathcal{H}$-matrix approximated calculation. It is obvious that the time consumption grows quadratic in relation to the number of degrees of freedom for the direct evaluation of the convolution integral and only proportional in relation to the number of degrees of freedom for the hierarchical matrix approximation.

Degrees of freedom	CPU time	
	direct calc.	$\mathcal{H}$-matrix approx.
$[-]$	$[s]$	$[s]$
246	75	32
756	631	245
2939	10396	992

Table 5.1. CPU time consumption for the direct evaluation and $\mathcal{H}$-matrix approximation of convolution integral (250th time-step)

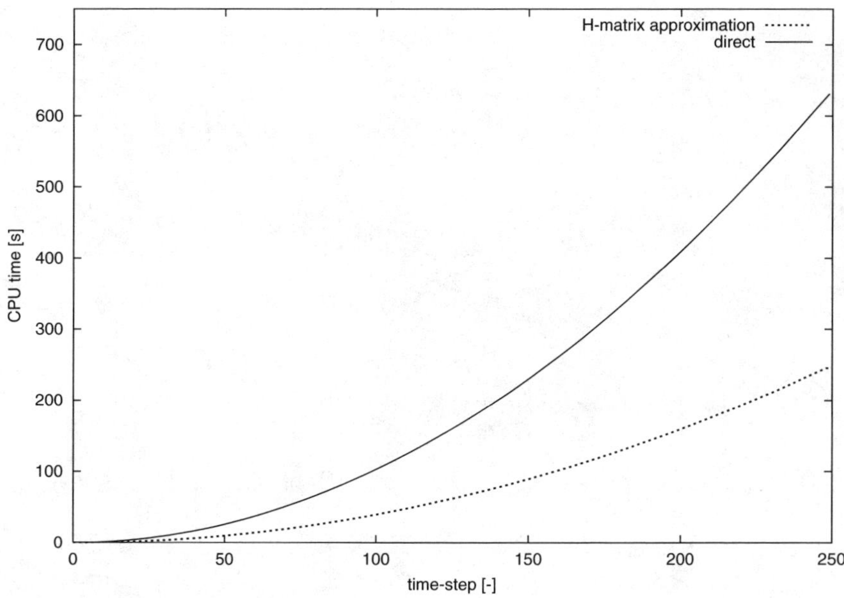

Fig. 5.14. CPU time consumption for direct calculation and $\mathcal{H}$-matrix approximation (sound insulating wall)

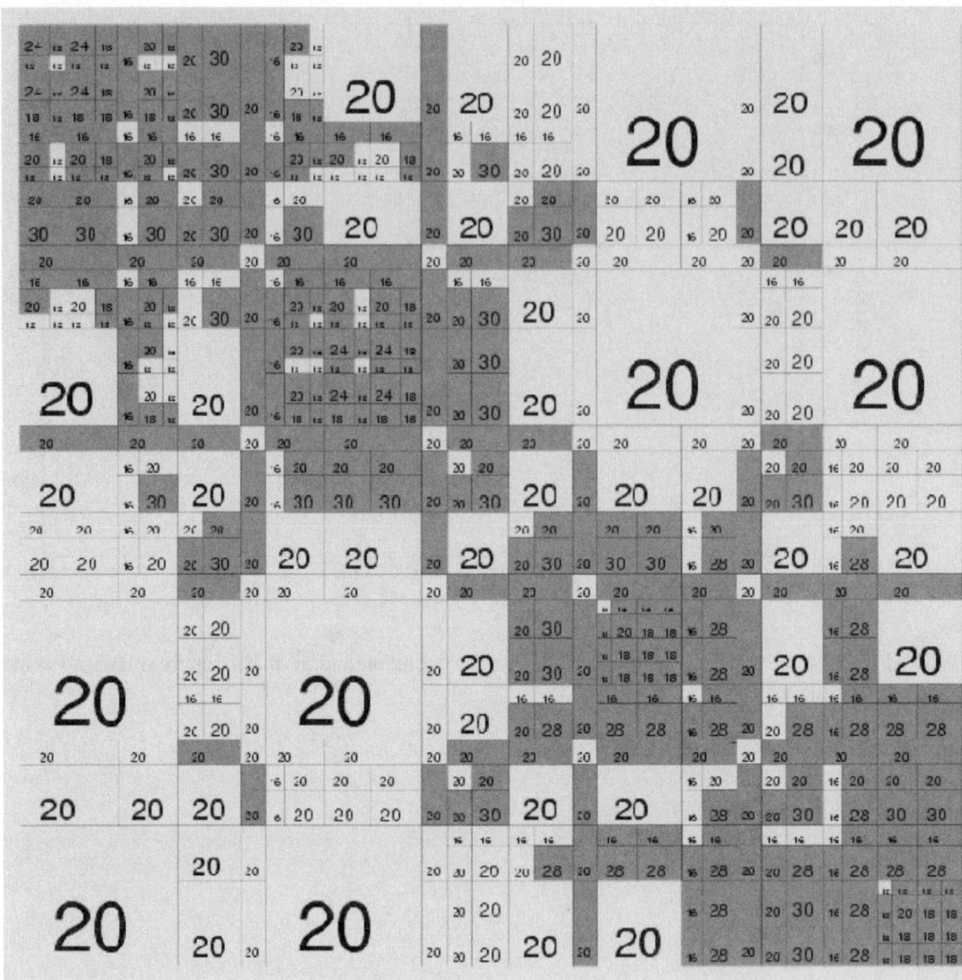

Fig. 5.15. Sound barrier example: $\mathcal{H}$-matrix structure of an influence matrix, rank of matrix block $k \leq 20$

5.4.3 Waveguide (time domain analysis)

As a last example for a wave propagation problem, a waveguide filled with air with different numbers of openings (three model variants, see Fig. 5.16) is considered.

System

The waveguide is assumed to have acoustically rigid walls. Fig. 5.16 shows the geometrical setting and the boundary conditions for three different model variations. For all models a constant Neumann boundary condition ($1\frac{m}{s}$) at one side face of the interior domain is prescribed. Model (a) is a closed box, whereas models (b) and (c) are coupled with the infinite space, also consisting of air.

Discretisation

The fluid of the waveguide is discretised with $5 \times 5 \times 40 = 1000$ eight-node three-dimensional fluid elements. Each element has a length, width and height of $0.02m$. For model (a) no scaled boundary finite elements are applied, because this is an explicit interior problem. At the second geometry (b) the opposite face of the excited side of the waveguide is coupled with the infinite space, where $5 \times 5 = 25$ scaled boundary finite elements are necessary. At the third geometry named model (c), additionally, the ceiling is coupled with the infinite space. Here, 225 ($5 \times 40 + 5 \times 5$) SBFEs are necessary. For a finite element analysis the time-step size can be calulated with $\Delta t \leq \frac{\min \ell_e}{10c}$ (see Sec. 5.4.2) which results in $\Delta t \leq \frac{0.02m}{10 \cdot 346 \frac{m}{s}} = 5.8 \cdot 10^{-6}s$. For a scaled boudary finite element analysis, the time-step size is ruled by the smallest distance $\min d_e$ of scaled boundary finite elements to the scaling centre: $\Delta t \leq \frac{\min d_e}{15c}$. For model (c) $\min d_e = 0.05m \Rightarrow \Delta t \leq \frac{0.05m}{15 \cdot 346 \frac{m}{s}} = 9.6 \cdot 10^{-6}s$. Thus, a constant time-step $\Delta t = 5 \cdot 10^{-6}s$ is chosen for all calculations due to finite element limitations.

Results

Fig. 5.17 shows the calculated pressure at the centroid of the side-face with prescribed sound impact versus time for model (a). It is obvious that the analytical and numerical solution are nearly identical.

Fig. 5.18 shows the calculated pressure versus time for the beginning of the simulation ($t \leq 0.002s$) for all three models. Here, the influence of the infinite space is evident after a very short simulation time at model (c), because the open (or coupled with infinite space) ceiling is in direct contact with the excited wall, see Fig. 5.16.

Fig. 5.19 depicts the calculated sound pressure at the centroid of the side-face for model (c). Here, beside the direct solution, the result of a $\mathcal{H}$-matrix approximation is shown. Results of direct and $\mathcal{H}$-matrix calculations are almost the same. The maximal relative error between direct and $\mathcal{H}$-matrix calculation is 10^{-7}. The CPU time consumption for the $\mathcal{H}$-matrix evaluation of the convolution integral for 1500 time-steps is significantly reduced to 41.8% of the direct calculation (see Fig. 5.20).

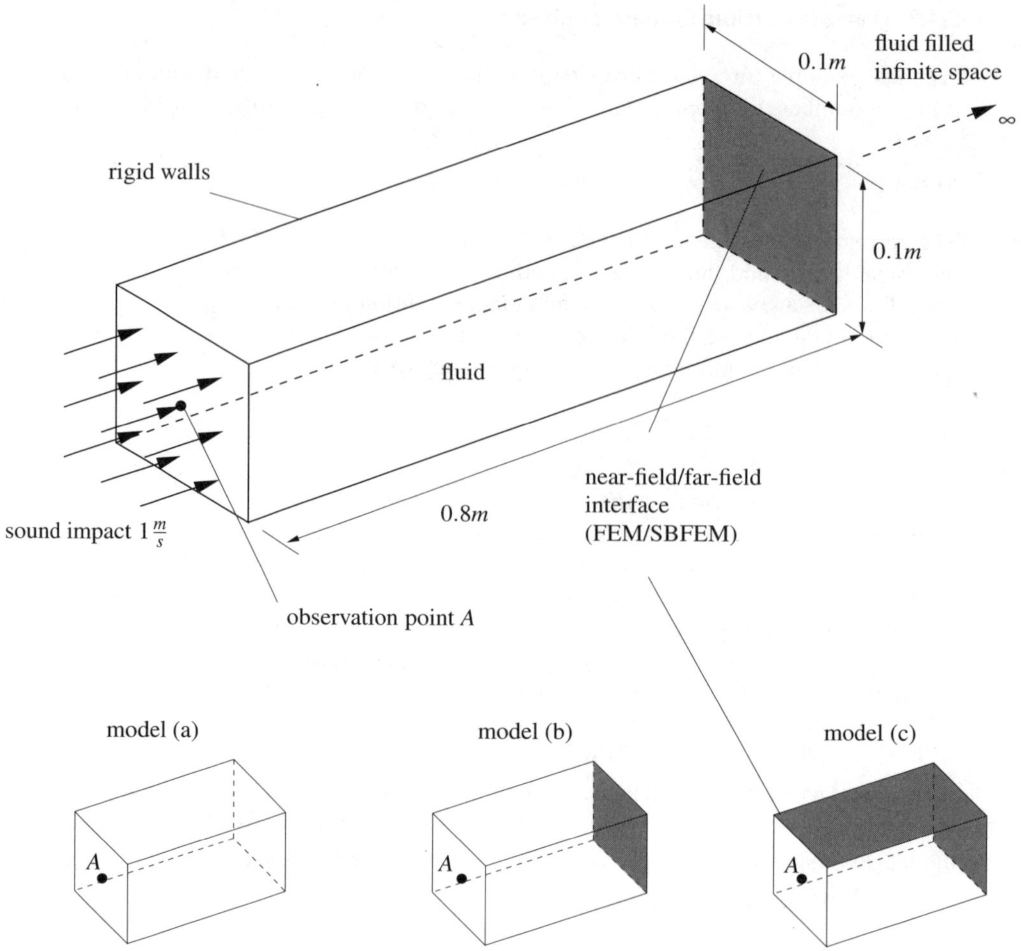

Fig. 5.16. Geometrical settings and boundary conditions of the waveguide

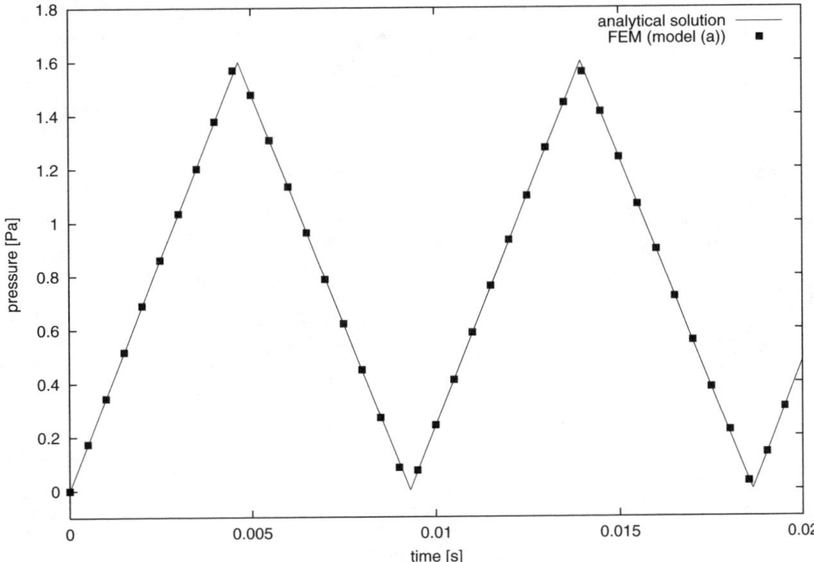

Fig. 5.17. Sound pressure at centroid of excited side-face versus time: model (a) (analytical and FE solution), $t \leq 0.02s$

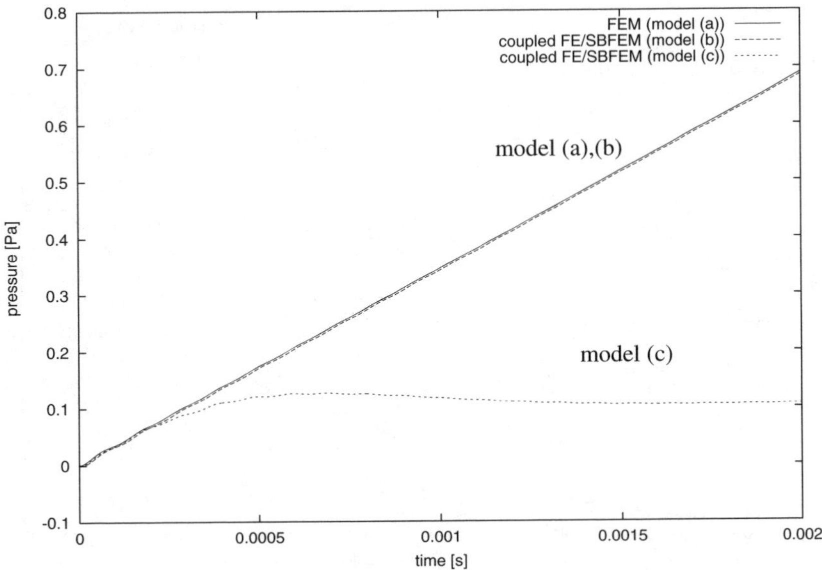

Fig. 5.18. Sound pressure at centroid of excited side-face versus time: model (a), (b) and (c), $t \leq 0.002s$

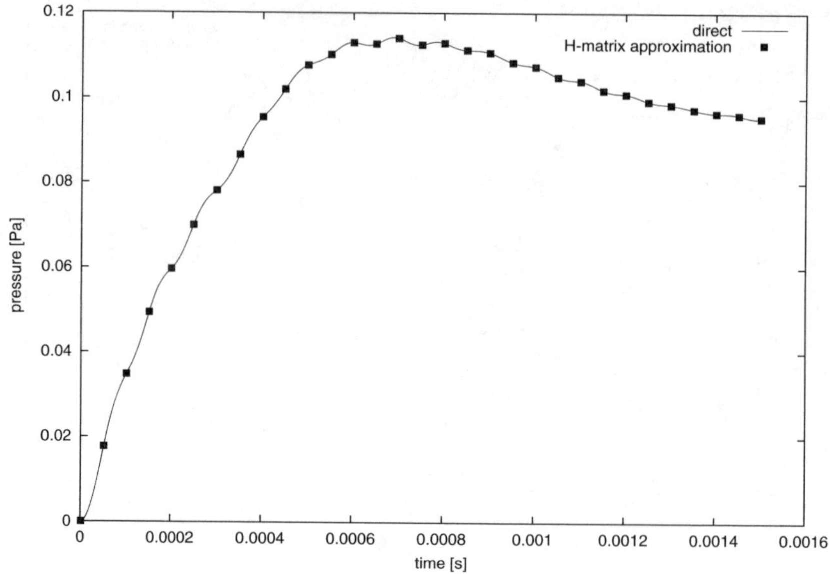

Fig. 5.19. Sound pressure at centroid of excited side-face versus time: model (c) (direct and $\mathcal{H}$-matrices solution)

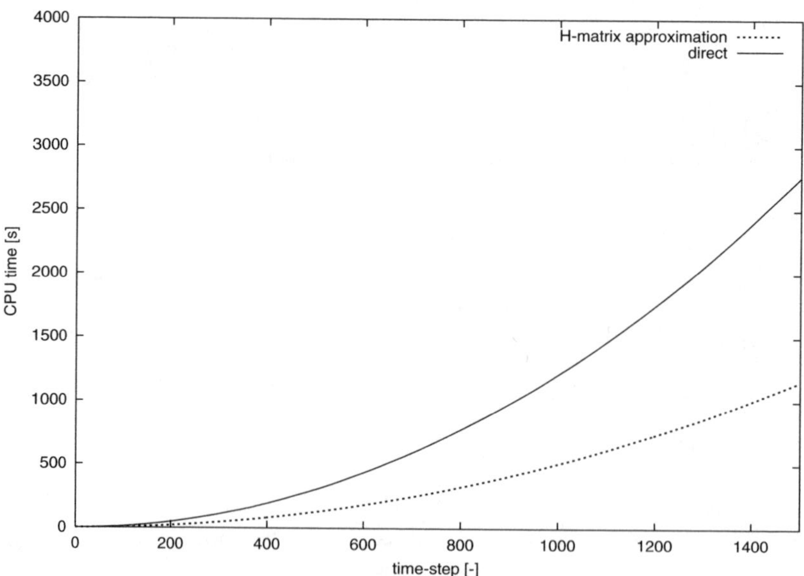

Fig. 5.20. CPU time consumption versus time-step: model (c) (direct and $\mathcal{H}$-matrices solution)

5.5 Concluding remarks

Comparison of the coupled FE/SBFE method with the BEM or coupled FE/BE method of the first and second example shows that the results are close. The calculations are performed in frequency and time domain. Here, the advantage of the presented approach is (as in the vector wave simulation) that the SBFEM is easy to incorporate in an existing FE code, i.e., the coupling of two-dimensional scaled boundary finite elements with three-dimensional fluid finite elements is rather trivial. Furthermore, at the SBFEM the results of the calculated influence of the infinite far-field is stored, so different load cases and geometries of the near-field can be simulated with this method without high costs.

In the third example, a three-dimensional coupled FE/SBFE simulation in the time domain as application of $\mathcal{H}$-matrices for the evaluation of the involved convolution integral is presented. This approximation reduces the computational effort from quadratic to linear dependency on degrees of freedom, without loosing accuracy. Since convolution integrals arise in various large-scale engineering computations the application of $\mathcal{H}$-Matrix technique might turn out to be a highly promising way to pursue.

6. Offshore wind energy conversion systems

In present times, the utilisation of regenerative energy becomes more and more important all over the world. Additionally, the process of producing energy should have negligible or small environmental impacts. The offshore wind energy is one possibility to support such energy – providing quite stable energy with small environmental impact.

The construction of wind turbines has been developed in the last decades especially in the countries of the European Community [98]. Starting with rather small turbines on land, these structures became very large (rotor diameter up to $120m$, and hub height up to $100m$) and placed offshore because of better wind conditions.

The tower and rotor systems of offshore turbines or offshore wind energy conversion systems (OWECS) are often slightly modified components that were originally designed for onshore use, while foundations of offshore turbines are special constructions (see Sec. 6.5.1). There are some tower and rotor system designs, however, that make use of the unique offshore environment. For example, floating turbines do not require an expensive foundation for deep water sites, or offshore turbines could be used to generate hydrogen through electrolysis of seawater [3].

The offshore environment forms some engineering challenges, as it is quite corrosive and abrasive. Therefore OWECS are outfitted with corrosion protection measures like coatings and cathodic protection. Repairs and maintenance are much more difficult and much more costly than on onshore turbines.

Since the average wind speed is usually higher over open water, offshore turbines can use shorter towers. In stormy areas with extended shallow continental shelves – such as Denmark, Netherlands or the German Bight – turbines are reasonably easy to install and give good service. Denmark's wind generation, e.g., provides about 12% to 15% of total electricity demand in the country, with a high percentage of offshore wind farms. Denmark plans to increase contribution of wind energy to as much as half of its electrical supply [42].

6.1 Short history of wind conversion systems

Wind machines were used for grinding grain in Persia as early as 200 B.C. This type of machine spread throughout the Islamic world and was introduced by crusaders to Europe in the 13th century. Windmills have been used to grind grain since around the

10th century in Europe. By the 14th century Dutch windmills were in use to drain areas of the Rhine River delta.

Wind turbines, machines that harness the energy of the wind to generate electrical energy, were first invented in the 19th century in Denmark. By 1900 there were about 2500 windmills for mechanical loads such as pumps and mills, producing an estimated combined peak power of about $30MW$. The first windmill for electricity production was built in Denmark in 1890, and in 1908 there were 72 wind-driven electric generators from $5kW$ to $25kW$. The largest machines were on $24m$ towers with four-bladed $23m$ diameter rotors [136].

Until approximately 1930 windmills were mainly used to generate electricity on farms, mostly in the USA where distribution systems had not yet been installed. In this period, high tensile steel was cheap, and windmills were placed atop prefabricated open steel lattice towers. A forerunner of modern horizontal-axis wind generators was in service at Yalta, USSR in 1931 [134]. This was a $100kW$ generator on a $30m$ tower, connected to the local $6.3kV$ distribution system. It was reported to have an annual load factor of 32%, not much different from current wind machines.

The wind energy industry as it is known today, really began in the early seventies, when the oil price shock caused many governments around the world to implement research programmes to investigate power generation technology from non-oil dependent sources, including wind energy. However, once the oil price stabilised, these research programmes were curtailed and no commercial plants were in fact built in this short period. The second oil price shock in the late seventies revitalised the efforts and led to commercial developments, such as large fields of wind turbines in California.

The first ideas for generating electricity using wind in the offshore environment came up in the late seventies and numerous feasibility studies where undertaken in the following decade. It was only in the early nineties that the first prototype offshore wind farms were actually built: a single turbine at Nogersund, Sweden, 1990 [79], and at Vindeby in Denmark in 1991 [80]. The performance of the wind farm at Vindeby has been heavily monitored and evaluated. Much useful knowledge has been gained on the performance of wind turbines offshore but also on the performance of wind turbines in general, specifically because of the low turbulence conditions. Vindeby is located in a relatively protected part of the Baltic Sea, surrounded by islands but even there access has been a problem because of excessively high waves. The Vineby wind farm consisted of 11 wind turbines, rated at $450kW$, giving a total farm output off almost $5MW$; by way of illustrating the progress made in this industry, the largest single prototype wind turbines currently being installed have the same rated power as that whole wind farm.

Vindeby was soon joined by a second wind farm in Denmark, at Tuno Knob [105], and two further wind farms were built in the Netherlands [145]. These were built on monopiles, which is becoming the preferred support structure for offshore wind turbines. In 1995 to 2000 there was less activity, with a single wind farm being built at Bockstigen-Valor in Sweden. In the year 2001 new wind farms were constructed using large multi-megawatt wind turbines; the largest is that of Middelgrunden, Denmark. It consists of 20 turbines each rated $2MW$ and is located on concrete gravity bases. The wind farm at Blyth, U.K., consists of only two turbines, again each of $2MW$. However they are located in the harshest environment yet: the north east coast of England facing directly into the North Sea, [4]. Two wind farms have also been built in Swedish waters: at Utgrunden (in 2000) and Yttre Stengrund (in 2001) [105].

6.2 Wind loading

Wind is continuously changing with time by nature. The wind has generally a higher annual average wind speed offshore than onshore, therefore energy capture and wind loading of offshore wind turbines differ from onshore turbines. Key parameters are the annual mean wind speed and the extreme gust wind speed. In the offshore environment, winds are generally characterised not only by higher annual average and extreme mean wind speeds, but also by lower turbulence and gust factors along with a smoother wind shear and more stable wind directionality. The annual mean wind speed determines the energy available per unit of swept area, while the extreme gust wind speed defines the sizing of the main structural compliances of the wind turbine, such as tower and foundation . For most sites this offers opportunities for design optimisation. At exposed sites these good natured effects are offset by both the more frequent high wind speeds, which dominate the fatigue, and stronger extreme gusts governing limit state.

To perform reliable wind turbine loadings and energy capture calculations it is also necessary to consider the mean wind speed profile and wind turbulence.

6.2.1 Wind speed distribution and wind profile

Not only the average wind speed (see Fig. 6.1), but also the wind speed distribution differs at onshore and offshore locations. At sea the shape of the probability density function of the mean wind speed is shifted to higher wind speeds corresponding to a larger Weibull shape factor [86]. Generally, on sea the annual average wind speed and the shape factor are larger. Both properties result in a larger probability of high wind speeds and favour a high specific rating for OWECS in terms of rated power per unit swept area. This can be realised for instance by a greater generator capacity at the same rotor diameter compared to onshore siting.

Fig. 6.1 (after [86]) illustrates how the wind profile develops at two typical onshore and offshore sites. The hub height at offshore sites is commonly chosen lower since relatively little extra energy can be gained by a taller structure, which is then relatively more expensive.

A common representation of a wind shear profile with a firm physical basis is the Prandtl logarithmic law model [107]:

$$\dot{x}_z = \frac{\dot{x}_{fr}}{k} \ln \left(\frac{z}{z_0} \right) , \qquad (6.1)$$

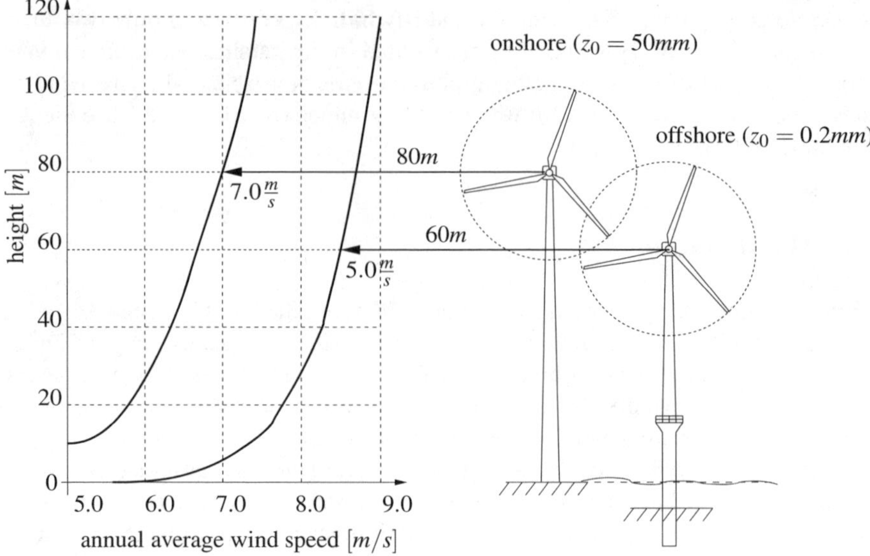

Fig. 6.1. Logarithmic wind profiles for typical onshore and offshore sites

where $\dot{x}_z$ is the mean wind speed at height z, k the von Karman constant ($k \approx 0.4$), z_0 is the surface roughness length and $\dot{x}_{fr}$ is the friction velocity , with

$$\dot{x}_{fr} = \frac{\dot{x}_{10}}{2.5\ln(10m/z_0)} \cdot \tag{6.2}$$

Equation (6.1) is only valid for the surface layer which extends to approximately $100m$ height. An empirical modification to this equation extends its range to heights of about $300m$ (see [73]):

$$\dot{x}_w(z) = \frac{\dot{x}_{fr}}{k} \left[\ln\left(\frac{z}{z_0}\right) + 5.75\left(\frac{z}{h}\right) \right] , \tag{6.3}$$

where the height of the atmospheric boundary layer h is given by:

$$h = \frac{\dot{x}_{fr}}{6f_C} , \tag{6.4}$$

where $f_C = 2\Omega\sin(\phi)$ means the Coriolis parameter , $\Omega = 7.29 \cdot 10^{-5}\frac{rad}{s}$ is the angular speed of the earth , and ϕ is the angle of latitude of the site under consideration.

For the open sea, z_0 is depending upon the sea state, and, therefore, on the mean wind speed. The roughness length z_0 is not directly related to the wave height, but rather to small wavelets or ripples on the surface of the waves. A commonly used expression is that one due to Charnock [35]:

$$z_0 = \frac{\dot{x}_{fr}^2}{gc} , \tag{6.5}$$

where g is the gravity constant and the dimensionless constant $c \approx 60$. Equation (6.5) needs to be solved iteratively for the mean wind speed at $10m$ height ($\dot{x}_{10}$) and the surface roughness z_0. The reduced surface roughness length z_0 at sea results in a steeper vertical wind profile (see Fig. 6.1) and a lower turbulence intensity. Charnock proposed a constant ratio between skin friction velocity and roughness, valid only for the equilibrium state between wind and wind generated waves [35]. However, a generally accepted mean value of roughness length for sea conditions is $2 \cdot 10^{-4}m$ [138], compared to $0.01m$ - $0.25m$ for sites on land [54]. In the case of waves caused by winds acting over a short length of water (a short fetch), or from a recently developed wind field, c is no longer a constant, but depends on the wave age.

For engineering purpose the wind speed profile is expressed by equation (6.1) or by a power law, where the wind speed is a function of non-dimensional height with the wind shear exponent as a parameter [57]. Both relations are only valid in the case of neutral atmospheric stability where thermally induced turbulences are negligible compared to mechanically induced turbulences, and thus there is no vertical heat exchange [98]. Although other conditions of atmospheric stability may be modelled by means of a diabatic wind profile. There is evidence to suggest that for the relatively high wind speeds needed for wind energy applications (OWECS), an assumption of neutral stability is adequate [48].

6.2.2 Atmospheric turbulence

Turbulent gusts are caused by eddies in the wind flow. For the wind field simulation it is assumed that they travel with the mean wind speed according to the commonly applied frozen flow hypothesis (Taylor [130]). Eddies have different frequencies, amplitudes and phases and may intersect each other. The characteristic length of turbulence structures, denoted as integral length scales, is of same magnitude as the rotor diameter of large wind turbines. The rotational period is commonly smaller than the time for a turbulence bubble to pass the swept area. Consequently most of the gusts affect only parts of the rotor but are seen several times by blades. This "rotational sampling" or "eddy slicing" transforms the wind spectrum seen by a rotating observer leading to a lumping of excitation energy at the rotor frequency and multiples of this frequency [86].

The temporal and spatial turbulence properties can be represented by power spectra associated with the three-dimensional velocity vector and a coherence function accounting for the spatial correlation. Auto and cross spectral density functions can be determined at discrete points of the swept area [146] or on the rotor blades [45], both based on the works of Shinozuka [118] and Veers [139]. Random representations in the time domain are then generated by the inverse fast Fourier transformation.

For design load calculations of wind turbines, turbulent wind speed variations are generally considered only in the longitudinal direction. The intensity of turbulence $I_u(z)$ is defined as the ratio of the standard deviation of wind speed variations to mean wind speed, both calculated or measured at same height z.

Mean and standard deviation quantities are specified with respect to an averaging period of one hour. According to the Engineering Sciences Data Unit International (ESDU) [48] the longitudinal turbulence intensity at height z is given by:

$$I_u(z) = \frac{0.867 + 0.556 \log_{10}(z) - 0.246[\log_{10}(z)]^2}{\ln(z/z_0)} . \tag{6.6}$$

As mentioned above, the roughness length z_0 depends on the mean wind speed. The turbulence intensity of flow over the sea is therefore also a function of wind speed. Offshore wind measurements show that the turbulence intensity fluctuates with the mean wind speed [36]. The mean turbulence intensity generally increases with the mean wind speed, as the roughness of the sea surface is a function of the mean wind speed. However, measurements in the Danish North Sea have shown that at low wind speeds the turbulence intensity can be high, it decreases with the mean wind speed up to about $10\frac{m}{s}$ and then increases again with the mean wind speed [71].

The measured turbulence intensity varies also for a given mean wind speed. The spreading of the turbulence intensity is smaller at high wind speeds and larger at low wind speeds. Generally, it is assumed that the offshore wind has lower turbulence intensity than the onshore wind. Measurements show average values of the turbulence intensity of around 10% [36]. The offshore wind turbine guideline from Germanischer Lloyd [57] prescribes a turbulence intensity of 12%. For a turbine in a wind farm, the level of turbulence can increase due to the wake effect up to 20%.

6.2.3 Spectral characteristics

Information concerning the spectral characteristics of offshore wind is rare. The information which is available in literature stems mainly from measurements at offshore structures outfitted with meteorological equipment.

In the view of scarcity of turbulence measurements for offshore wind a standard spectral model of turbulence is used for the numerical simulation in this work. Although standard spectral models may under-estimate low frequency turbulent energy in the wind over sea, it is considered that their use is appropriate for the numerical model in Sec. 6.6. Here, the turbulence model due to von Karman has been chosen. The von Karman spectral equations are generally accepted to be the best analytical representation of isotropic turbulence and are recommended by ESDU [48]. Use of the von Karman model is also consistent with the expressions stated in (6.1) and (6.6) for the logarithmic shear profile and turbulence intensity, respectively. The von Karman model defines the auto spectral density of longitudinal wind speed variations as follows:

$$\frac{f S_{uu}(f)}{\sigma_w^2} = \frac{4\tilde{f}}{(1 + 70.8\tilde{f}^2)^{5/6}} , \tag{6.7}$$

with S_{uu} is the auto spectral density of wind speed, σ_w is the standard deviation of wind speed $\dot{x}_w$, f the frequency $[Hz]$ and $\tilde{f}$ is the non-dimensional frequency, given by:

$$\tilde{f} = \frac{L_u f}{\dot{x}_w} , \tag{6.8}$$

where L_u is the internal scale length of turbulence . The turbulence length scale is dependent on the surface roughness and according to ESDU [48] given by:

$$L_u = \frac{25z^{0.35}}{z_0^{0.0063}} \cdot \tag{6.9}$$

A model of a turbulent wind field suitable for wind turbine loading calculations requires a correct representation of both the temporal and spatial structure of the longitudinal wind speed fluctuations. The cross correlation function relating wind speed fluctuations at two points separated in space can be obtained from the von Karman equations for isotropic turbulence by assumption of Taylor's frozen flow hypothesis [130]. The resulting expression is given in a technical report of the ESDU [48].

6.3 Wave- and sea ice loading

To evaluate the fluid-induced forces acting on the foundation or part of the tower which is below the water level of the OWECS it is necessary to know its surrounding hydrodynamic flow field. For offshore structures, the flow field arises from time-varying natural processes: winds, currents, and surface gravity waves. In the following subsection, focus is laid on gravity waves, which are usually wind-generated.

It has been observed that these time-dependent waves occur on two different time scales:

- The shorter time scale, measured in minutes or seconds, is useful for describing detailed features such as wind guts and surface periods. This shorter scale corresponds most closely to the response time of fixed offshore structures.
- The longer time scale, measured in hours, days or years, is useful for describing variations in wave intensity and in its statistics. This time scale is also important because OWECS may fail in low cycle fatigue fracture after months or years of service, see Etube [52].

There are two fundamentally different descriptions of surface gravity waves: deterministic and probabilistic. Deterministic descriptions, analytical or numerical, are used to characterise the short time scale feature of waves. In both descriptions, the linear wave theory is important for two reasons: it is simple to apply when estimating forces on offshore structures during the preliminary phases of design; and it affords a simple basis for estimating the probability of failure of a given structure.

In the statistical description of offshore waves, irregular waves based on experimental measures of surface wave heights are analysed in a statistical sense. With this method, it is possible to obtain wave spectra and other statistical parameters which will prove useful in predicting the response of offshore structures to irregular waves.

6.3.1 Deterministic description of waves with the linear wave theory

Water kinematics can be determined by the linear wave theory. The linear wave theory is alternatively known as Airy's theory , small amplitude theory, and first order theory, developed by G.B. Airy in 1845 [2]. It gives a relatively simple solution for periodic gravity waves. These waves are assumed to have a constant wave length λ, constant wave height H, water of constant depth d and propagating in a constant direction x_1, see Fig. 6.2. Because the waves are propagating in one direction, the flow can be described in two dimensions $\mathbf{x} = (x_1; x_2)$ where x_2 is measured vertically upward from still water level (SWL). The water elevation $\eta(x_1; t)$ is also measured upward from SWL. The water is assumed to be incompressible and the flow to be inviscid and irrotational. The linear wave theory is constrained to small amplitude waves which allows the boundary conditions to be linearised. Nevertheless the linear wave theory is appropriate for estimating the fatigue life time of the structure, since small but frequently occurring waves form the major part of the loading.

Boundary conditions for the bottom and the free surface are necessary to solve the Laplace equation for the velocity potential. The bottom boundary condition and kinematic free surface boundary condition require that there is no flow through sea bed and free surface. The dynamic free surface boundary condition requires that the pressure at the water surface is equal to the atmospheric pressure.

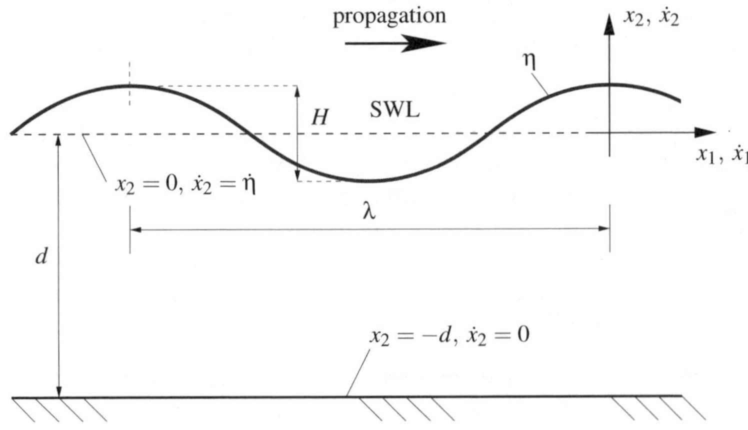

Fig. 6.2. Linear wave theory: schematic sketch of wave propagation

The inherent assumptions of the linear wave theory are in detail:

- amplitude a of surface disturbance is very small relative to wave length λ and water depth d,
- water depth d is uniform,
- the fluid is non-viscous, irrotational, incompressible and homogeneous,
- Coriolis forces due to earth's rotation are negligible,

- the surface tension is negligible,
- sea floor is smooth and impermeable, and
- the sea level atmospheric pressure p_a is uniform.

The hydrostatic pressure p_h is given by:

$$p_h = -\rho_f g (d - x_2) \,, \tag{6.10}$$

with ρ_f is the density of the fluid . In the following, the dynamic water pressure is denoted by p.

Governing equation is the two-dimensional Laplace equation for an incompressible fluid:

$$\Delta \Phi = 0 \ \ \text{with} \ \ \Phi = \Phi(x_1, x_2) \,, \tag{6.11}$$

where Φ is the velocity potential:

$$\dot{\mathbf{x}} = \nabla \Phi \,. \tag{6.12}$$

With the assumptions mentioned above, the following conditions can be mathematically formulated:

The *bottom boundary condition:*

$$\Phi_{,2} = 0 \ \ \text{at} \ \ x_2 = -d \,, \tag{6.13}$$

the *kinematic free surface boundary condition:*

$$\dot{x}_2 = \dot{\eta} \ \ \text{at} \ \ x_2 = 0 \,, \tag{6.14}$$

which is equivalent to:

$$\Phi_{,2} = \dot{\eta} \ \ \text{at} \ \ x_2 = 0 \,, \tag{6.15}$$

and the *dynamic free surface boundary condition:*

$$\dot{\Phi} = -g\eta \ \ \text{at} \ \ x_2 = 0 \,. \tag{6.16}$$

Combining (6.15) and (6.16) results in the boundary condition at SWL ($x_2 = 0$):

$$\ddot{\Phi} + g\,\Phi_{,2} = 0 \,. \tag{6.17}$$

With these boundary conditions, the solution $\Phi(x_1, x_2, t)$ of the Laplace equation in two dimensions (6.11) is (see, i.e. [144]):

$$\Phi(x_1, x_2, t) = \frac{a\omega}{k} \frac{\cosh[k(x_2 + d)]}{\sinh kd} \sin(kx_1 - \omega t) \,. \tag{6.18}$$

In (6.18) ω is the wave frequency and k defines the wave number , expressed independency of wave length λ with:

$$k = \frac{2\pi}{\lambda} \,. \tag{6.19}$$

From (6.16) and (6.18) the water elevation or surface wave profile $\eta(x_1,t)$ can be obtained:

$$\eta(x_1,t) = ae^{-i(kx_1-\omega t)}. \tag{6.20}$$

The resulting velocities and accelerations of the fluid in x_1- and x_2-direction are:

$$\dot{x}_1 = a\omega\frac{\cosh[k(x_2+d)]}{\sinh kd}\cos(kx_1-\omega t), \tag{6.21}$$

$$\dot{x}_2 = a\omega\frac{\sinh[k(x_2+d)]}{\sinh kd}\sin(kx_1-\omega t), \tag{6.22}$$

$$\ddot{x}_1 = a\omega^2\frac{\cosh[k(x_2+d)]}{\sinh kd}\sin(kx_1-\omega t) \text{ and} \tag{6.23}$$

$$\ddot{x}_2 = -a\omega^2\frac{\sinh[k(x_2+d)]}{\sinh kd}\cos(kx_1-\omega t). \tag{6.24}$$

The dispersion relation gives the connection between the wave frequency ω, the wave number k and the water depth d. It results also from the given boundary conditions:

$$\omega^2 = gk\tanh kd. \tag{6.25}$$

For the assumption of deep water, i.e., the product kd is sufficient large, $\tanh kd \approx 1$. For $d > \lambda$ the maximum error ε is less than 0.0007%:

$$d > \lambda \to \tanh kd > \tanh 2\pi = 0.999993 \to \varepsilon < 0.0007\%,$$

and for $d > \frac{\lambda}{2}$ the maximum error ε is less than 0.37%:

$$d > \lambda/2 \to \tanh kd > \tanh\pi = 0.9963 \to \varepsilon < 0.37\%.$$

In the literature, i.e., [109, 34], the assumption of deep water is usually equal to a water depth $d > \frac{\lambda}{2}$ ($\varepsilon < 0.37\%$). For the assumption of deep water, i.e., $\tanh kd$ assumed to be one, the equation (6.18) reduces to:

$$\Phi(x_1,x_2,t) = \frac{a\omega}{k}e^{kx_2}\sin(kx_1-\omega t). \tag{6.26}$$

The resulting simplified velocities and accelerations are:

$$\dot{x}_1 = a\omega e^{kx_2}\cos(kx_1-\omega t), \tag{6.27}$$

$$\dot{x}_2 = a\omega e^{kx_2}\sin(kx_1-\omega t), \tag{6.28}$$

$$\ddot{x}_1 = a\omega^2 e^{kx_2}\sin(kx_1-\omega t), \tag{6.29}$$

$$\ddot{x}_2 = -a\omega^2 e^{kx_2}\cos(kx_1-\omega t). \tag{6.30}$$

The water elevation $\eta(x_1,t)$ remains the same for deep water assumption. Finally, the dispersion relation (6.25) reduces to:

$$\omega^2 = gk. \tag{6.31}$$

The water particle velocities $\dot{x}_1,\dot{x}_2$ and accelerations $\ddot{x}_1,\ddot{x}_2$ are those used in Morison's equation (see Sec. 6.3.3) to compute the drag and inertial forces for these waves on offshore structures.

6.3.2 Statistical description of offshore waves

The linear wave theory describes sea waves as a harmonic function with fixed frequency at a constant depth. In order to get a more realistic idea of the wave forces on offshore structures, the sea surface elevation should be considered as a time dependent *random process* . If the surface elevation is measured at different points in time and several records are obtained then the random process is represented by an ensemble of sample functions of every record.

A random process is defined as Gaussian when all probability density functions of the sample functions are Gaussian. Such a random process is called stationary if the probability density functions and the statistical properties as mean, mean square and variance obtained for the sample functions do not depend on the chosen interval of time. In addition a stationary process is called an ergodic process if the probability density function is the same for each of the records [103]. The most engineering problems can be assumed to be stationary and ergodic. This leads to the advantage that all statistical properties of a process can be obtained from a single record.

The water surface elevation $\eta(t)$ (now calculated for a fixed point x_1, therefore $\eta = \eta(t)$) is a random process which can be represented by the associated, empirically derived spectral density function $S_{\eta\eta}(\omega)$. The spectral density function is an important statistical characteristic of the sea waves since it describes the energy distribution over the frequency range of every sea state.

To estimate a spectral density function which corresponds to the wave environment, wave records from that location need to be examined and certain statistical parameters to be extracted. Two statistical parameters which are representative of the wave records are the significant wave height H_s and the zero-crossing period T_0. H_s is the average of the highest third of wave heights, and the zero-crossing period T_0 is the average time between successive zero up-crossings.

Empirically derived *sea spectra* refer to a fully developed or duration limited sea. When the duration of the wind is insufficient for the lower frequency waves to develop the resulting spectrum will consist of mostly high frequency components: the sea is then said to be duration limited. When the point of observation is far from the coast and the wind has blown for a long time low frequency waves have developed and the spectrum corresponds to a fully developed sea.

The most commonly used spectrum for fully developed sea is the *Pierson-Moskowitz spectrum* [111]. Here, the input parameters are H_s and T_0. These two statistical parameters are obtained from a wave scatter diagram which gives information about the different sea states occurring during a period of time, usually one year. The random function of the water surface elevation is subsequently generated from the wave spectrum by Fourier series.

Another wave spectrum, especially developed for the North Sea, is the *Joint North Sea Wave Project (JONSWAP) spectrum* which modifies the Pierson-Moskowitz spectrum with a peak enhancement factor γ.

Pierson-Moskowitz sea spectrum

The Pierson-Moskowitz spectrum is a commonly used spectral density function for a fully developed sea: fetch and duration are considered infinite. For the applicability of such a model, the wind has to blow over a large area at nearly constant speed for many hours prior to the time when the wave record is obtained, and the wind should not change its direction significantly [34]. Despite these assumptions, this spectral model is applied extensively in the design of offshore structures and is one of the most representative models for waters all over the world [98].

The Pierson-Moskowitz spectrum in dependence of a peak frequency ω_m is given by [111]:

$$S_{\eta\eta}(\omega) = \frac{\alpha g^2}{\omega^5} \exp\left[-\frac{4}{5}\left(\frac{\omega_m}{\omega}\right)^4\right],$$
(6.32)

or, in dependency of the mean wind speed $\dot{x}_{19.5}$ at $19.5m$ above SWL:

$$S_{\eta\eta}(\omega) = \frac{\alpha g^2}{\omega^5} \exp\left[-\beta\left(\frac{g}{\omega \dot{x}_{19.5}}\right)^4\right],$$
(6.33)

with α and β given by:

$$\alpha = 4\pi^3\left(\frac{H_s}{gT_0^2}\right)^2 \quad \text{and}$$
(6.34)

$$\beta = 16\pi^3\left(\frac{\dot{x}_{19.5}}{gT_0}\right)^4.$$
(6.35)

For the North Sea these constants are estimated as $\alpha = 0.0081$ and $\beta = 0.74$. Equations (6.34) and (6.35) after Neumann and Pierson [102] give the empirical relationship[1] between the sea state parameters H_s and T_0 and the mean wind speed $\dot{x}_{19.5}$:

$$H_s = \frac{0.21}{g}\dot{x}_{19.5}^2,$$
(6.36)

$$T_0 = 0.81\left(\frac{2\pi}{g}\dot{x}_{19.5}\right).$$
(6.37)

The Pierson-Moskowitz spectrum can be expressed directly by H_s and T_0 using (6.36) and (6.37):

$$S_{\eta\eta}(\omega) = \frac{4\pi^3 H_s^2}{T_0^4 \omega^5} \exp\left[-\frac{1}{\pi}\left(\frac{2\pi}{\omega T_0}\right)^4\right].$$
(6.38)

---

[1] This relationship is not conform to the recommended data of the International Towing Tank Conference (ITTC) [114] for the Pierson-Moskowitz spectrum, since ITTC gives larger significant wave heights for given wind speeds $\dot{x}_{19.5}$.

The frequency ω_m, corresponding to the maximum energy density, is for the Pierson-Moskowitz spectrum:

$$\omega_m = \left(\frac{4}{5}\beta\right)^{1/4} \frac{g}{\dot{x}_{19.5}} . \tag{6.39}$$

Joint North Sea Wave Project (JONSWAP) spectrum

This spectrum is assumed to be especially suitable for the North Sea, and does not represent a fully developed sea. It is a reasonable model for wind generated sea when $3.6H_s < T_p < 5H_s$ (with T_p is the peak period).

The JONSWAP spectrum takes into account the higher peak of the spectrum in a storm for the same total energy as compared to the Pierson-Moskowitz spectrum and also the occurrence of frequency shift of the spectrum maximum. It represents the conditions in a duration limited sea. The JONSWAP spectrum modifies the Pierson-Moskowitz spectrum by introducing a peak enhancement factor γ. It models the worst case during a typical storm. The JONSWAP spectrum (analogous to (6.32) in dependency of ω_m) is given by [74]:

$$S_{\eta\eta}(\omega) = \frac{\alpha g^2}{\omega^5} \exp\left[-\frac{5}{4}\left(\frac{\omega_m}{\omega}\right)^4\right] \gamma^{\exp\left[\frac{-(\omega-\omega_m)^2}{2\sigma^2 \omega_m^2}\right]} , \tag{6.40}$$

where α is the same parameter used in Pierson-Moskowitz spectrum, γ is the peak enhancement factor ($1 < \gamma < 7$ with mean value 3.3 for the North Sea). However, after the authors of the computer code "Wave Analysis for Fatigue and Oceanography" (WAFO-Group) [141] a more correct approach is to relate γ to H_s:

$$\gamma = \exp\left[3.484\{1 - 0.1975(0.036 - 0.0056T_p/\sqrt{H_s})T_p^4/H_s^2\}\right] . \tag{6.41}$$

The relation between peak period T_p and mean zero-upcrossing period T_0 may be approximated by:

$$T_0 = \frac{Tp}{1.30301 - 0.01698\gamma + \frac{0.12102}{\gamma}} . \tag{6.42}$$

Corresponding to WAFO, this parameterisation is based on qualitative considerations of deep water wave data from the North Sea.

In (6.40) $\sigma = \sigma_a$ is the left side width (for $\omega \leq \omega_m$) with average value of 0.07, and $\sigma = \sigma_b$ is the right side width (for $\omega > \omega_m$) with average value of 0.09.

Comparison of Pierson-Moskowitz and JONSWAP spectrum

For a wind speed $\dot{x}_{19.5} = 10.3\frac{m}{s}$ ($H_s = 3.1m$) and $\dot{x}_{19.5} = 20.6\frac{m}{s}$ ($H_s = 8.1m$) [2] the Pierson-Moskowitz and JONSWAP spectra (for $\gamma = 3.3$ and $\gamma = 7.0$) are plotted in

----

[2] ITTC recommended values for the Pierson-Moskowitz spectrum: significant wave heights and related wind speeds [114]

Fig. 6.3 and 6.4. Here, the peak circular frequency ω_m is calculated with (6.39) ($\beta = 0.74$ for a North Sea site): $\omega_m = 0.8354 \frac{rad}{s}$ and $\omega_{m_2} = 0.4177 \frac{rad}{s}$. It is shown that for the same frequency ω_m the JONSWAP spectra are similar to the Pierson-Moskowitz spectrum with a peak enhancement factor γ.

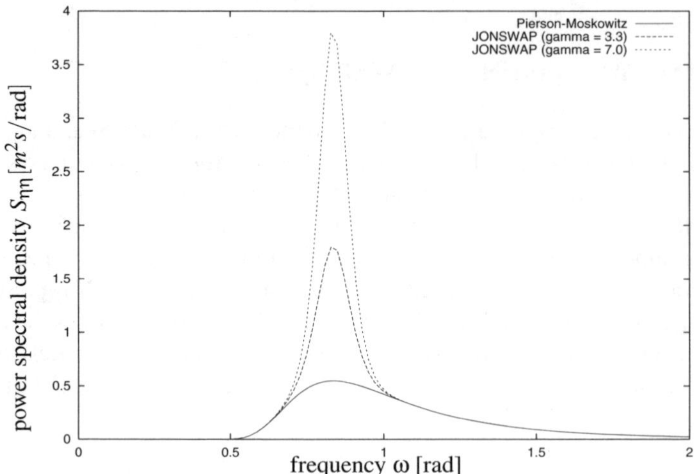

Fig. 6.3. Pierson-Moskowitz and JONSWAP spectra ($\gamma = 3.3$ and 7.0) for $\dot{x}_{19.5} = 10.3 \frac{m}{s}$ and $H_s = 3.1m$

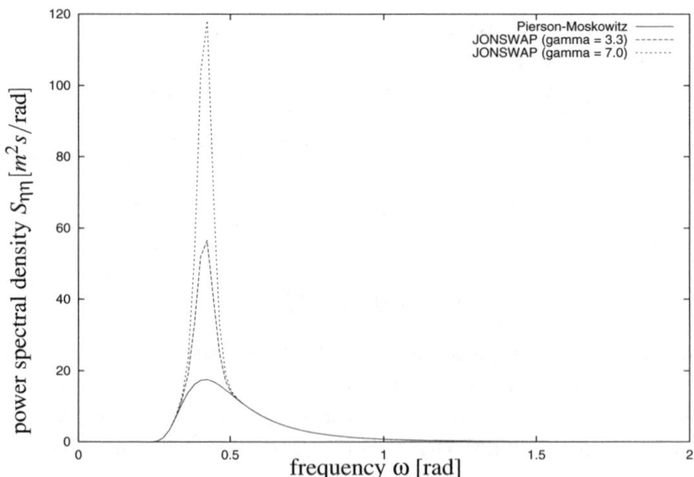

Fig. 6.4. Pierson-Moskowitz and JONSWAP spectra ($\gamma = 3.3$ and 7.0) for $\dot{x}_{19.5} = 20.6 \frac{m}{s}$ and $H_s = 8.1m$

If the peak frequency of the JONSWAP spectrum is assumed to be fetch dependent it can be calculated after Patel [109]:

$$\omega_m = \frac{2\pi 16.04}{(x_f \dot{x}_{10})^{0.38}} \, ,$$

(6.43)

where x_f is the fetch length . The wind speed $\dot{x}_{10}$ in $10m$ height can be calculated from the given wind speed in $19.5m$ height with:

$$\dot{x}_z = \frac{\dot{x}_{10}}{k \cdot 2.5 \ln\left(\frac{10m}{z_0}\right)} \ln\left(\frac{z}{z_0}\right) \, ,$$

(6.44)

where $z = 19.5m$, $z_0 = 2 \cdot 10^{-4}m$ (roughness length for offshore sites) and the von Karman constant $k = 0.4$. The wind speed in $10m$ height is then given by $\dot{x}_{10} = 0.942\dot{x}_{19.5}$, which results in $\dot{x}_{10_1} = 9.70\frac{m}{s}$ and $\dot{x}_{10_2} = 19.40\frac{m}{s}$.

Corresponding γ_i values are obtained from (6.41): $\gamma_1 = 3.24$, $\gamma_2 = 1.02$, $\gamma_3 = 9.16$ and $\gamma_4 = 2.49$. The value of $\gamma_3 = 9.16$ is out of bounds $1 < \gamma < 7$ for the JON-SWAP spectrum, because for a short fetch length of $25km$ the significant wave height $H_0 = 8.1m$ is not a realistic value for the North Sea. Therefore, γ_3 is set to 7, see Tab. 6.1, where additionally the resulting peak frequencies for fetch lengths of $25km$ and $50km$ are displayed. Fig. 6.5 depicts the Pierson-Moskowitz spectrum

i	x_f	$\dot{x}_{19.5}$	$\dot{x}_{10}$	ω_{m_i}	γ_i
$[-]$	$[km]$	$[m/s]$	$[m/s]$	$[rad/s]$	$[-]$
1	25	10.3	9.7	0.906	3.24
2	50	10.3	9.7	0.696	1.02
3	25	20.6	19.4	0.696	7.00
4	50	20.6	19.4	0.535	2.49

Table 6.1. Peak frequencies and peak enhancement factors for different fetch lengths and wind speeds

and JONSWAP spectra (fetch lengths $x_{f_1} = 25km$ and $x_{f_2} = 50km$) for a wind speed $\dot{x}_{19.5} = 10.3\frac{m}{s}$. The peak frequency for a fetch length of $50km$ (JONSWAP) is shifted to a lower frequency compared to the peak frequency of the Pierson-Moskowitz spectrum. For the smaller fetch length ($25km$) the peak frequency is shifted to a higher value. For both JONSWAP spectra the peaks are enhanced compared to the peak of the Pierson-Moskowitz spectrum.

Fig. 6.6 displays the Pierson-Moskowitz spectrum and JONSWAP spectra (fetch lengths $x_{f_1} = 25km$ and $x_{f_2} = 50km$) for a wind speed $\dot{x}_{19.5} = 20.6\frac{m}{s}$. It can be observed that for the fetch limited spectra the peaks are reduced and moved to higher frequencies.

Numerical generation of random waves

The random function of wave elevation can be generated from the power spectral density $S_{\eta\eta}(\omega)$ by a Fourier series. Thus, the wave elevation is expressed as a superposition of harmonic waves with amplitudes equal to the Fourier coefficients a_i:

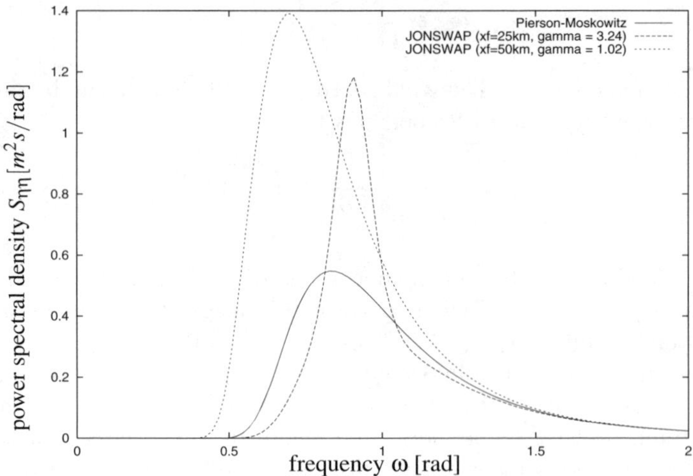

Fig. 6.5. Pierson-Moskowitz spectrum for $\dot{x}_{19.5} = 10.3\frac{m}{s}$ and JONSWAP spectra for $\dot{x}_{19.5} = 10.3\frac{m}{s}$ and $x_{f_1} = 25km$, $x_{f_2} = 50km$

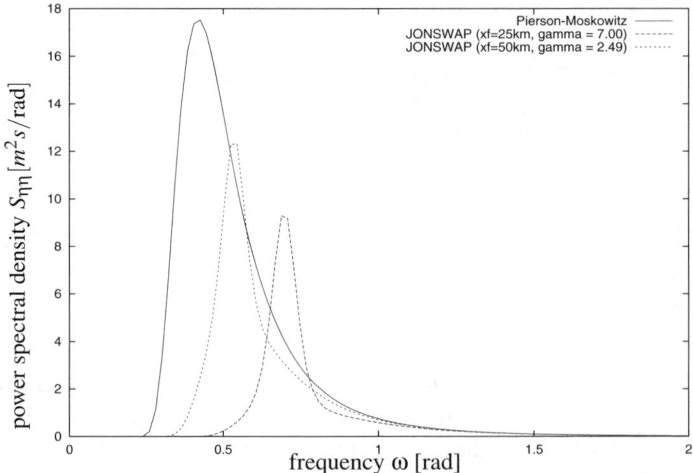

Fig. 6.6. Pierson-Moskowitz spectrum for $\dot{x}_{19.5} = 20.6\frac{m}{s}$ and JONSWAP spectra for $\dot{x}_{19.5} = 20.6\frac{m}{s}$ and $x_{f_1} = 25km$, $x_{f_2} = 50km$

$$\eta(t) \approx \sum_{i=1}^{N} a_i \cos(\omega_i t - \phi_i) \,, \tag{6.45}$$

with ω_i discrete angular frequency, and ϕ_i uniformly distributed phase angles, with $0 \leq \phi_i \leq 2\pi$. Assuming the linear wave theory, the wave elevation can be expressed as a function of time and horizontal position $x = x_1$:

$$\eta(x,t) \approx \sum_{i=1}^{N} a_i \cos(k_i x - \omega_i t - \phi_i) \,, \tag{6.46}$$

where the wave numbers k_i correspond to the frequencies ω_i of the harmonic waves via the dispersion relation (6.25) or the linearised form (6.31).

The power spectral density is the most important characteristic of the random function in the frequency domain since it contains the information about the distribution of the energy over the whole range of frequency. Fourier coefficients a_i can be calculated by:

$$a_i = \sqrt{2S_{\eta\eta}(\omega_i)\,d\omega_i} \,, \tag{6.47}$$

where $S_{\eta\eta}(\omega_i)$ is the value of the power spectral density at frequency ω_i, $d\omega_i$ is the frequency interval forming the regular wave of amplitude a_i. The factor two included in the expression for the amplitude a_i refers to the one-sided function of power spectral density. Fig. 6.7 illustrates the calculation of regular wave amplitudes where the area of the spectrum is subdivided into bars with width $d\omega_i$ (the constant frequency increment) and height equal to the value of the power spectral density at frequency ω_i: $S_{\eta\eta}(\omega_i)$.

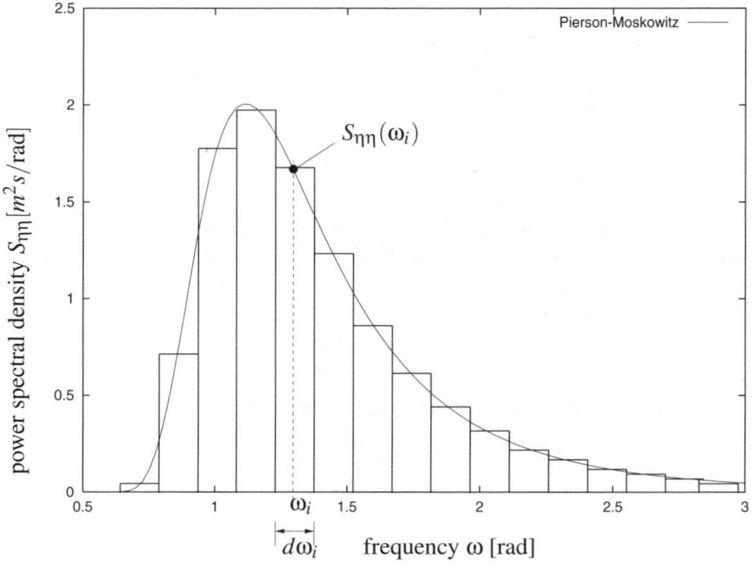

Fig. 6.7. Equidistant discretisation of a wave spectrum

6.3.3 Fluid-induced forces

In the following, an isolated, fully submerged, circular cylindrical solid for which the incident fluid velocity $\dot{x}_f$ is perpendicular to its longitudinal axis is assumed. In the plane flow case considered, shown in Fig. 6.8, the fluid or cylinder motion is in line with its net force per unit length q. Additionally, the fluid is assumed to be incompressible, and effects of nearby objects or solid boundaries are not included.

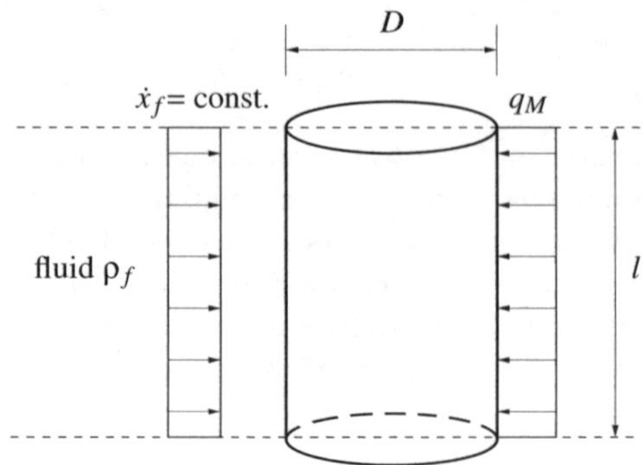

Fig. 6.8. A rigid cylinder in an ideal, accelerating fluid

Inviscid fluid flow

The inertia coefficient C_M relates the force per unit length q_M that is required to hold a rigid cylinder stationary in a fluid of uniform, constant free stream acceleration of magnitude $\ddot{x}_f$ (where the index f stands for fluid)[3]:

$$q_M = C_M \rho_f \pi r^2 \ddot{x}_f \,, \tag{6.48}$$

where ρ_f is the fluid density and r the cylinder radius ($D = 2r$).

$\frac{\ell}{D}$	C_M
1.2	1.62
1.5	1.78
5.0	1.90
9.0	1.96
∞	2.00

Table 6.2. Values of C_M in dependence of $\frac{\ell}{D}$

---

[3] For simplicity the index 1 at x_1 is dropped in the following, because of the one-dimensional description: $\mathbf{x} = x_1 = x$.

Tab. 6.2 shows some values of C_M for several ratios of cylinder length to diameter $\frac{\ell}{D}$. These results are reported by Wendel [142]. In contrast, after Kühn [86] a maximum value $C_M > 2$ is possible ($C_M \approx 2.1$ for $\frac{\ell}{D} = 10$).

Additionally, the movement of the structure has to be taken into account (also called water added mass force):

$$q_S = (C_M - 1)\rho_f \pi r^2 \ddot{x}_s , \tag{6.49}$$

where $\ddot{x}_s$ is the acceleration of the structure.

Viscous fluid flow

Measurements show that C_M is time-dependent, which is due to fluid viscosity. There is flow separation behind the cylinder, accompanied by differential pressure forces opposing cylinder motion [117]. Such forces are referred to as form drag. In applications, a root-mean-square (rms) measured average of the coefficient C_M is used. If such data is lacking, it is appropriate to choose $C_M = 2$ for design purpose, provided that the geometric ratio $\frac{\ell}{D}$ is much greater than one (see Tab. 6.2).

Another important parameter is the viscous, frictional drag coefficient, C_D. The force q_D per unit length can be defined as force, which is necessary to hold a fully immersed cylinder stationary as it is subjected to a constant free stream fluid velocity $\dot{x}_f$:

$$q_D = C_D \rho_f r |\dot{x}_f| \dot{x}_f . \tag{6.50}$$

For this flow case, the experimental relationships of C_D to two non-dimensional parameters, cylinder roughness and the Reynolds number , are well known [117]. Here, the Reynolds number Re is defined by:

$$Re = \frac{\rho_f \dot{x}_f D}{\mu_f} , \tag{6.51}$$

where μ_f is the absolute viscosity of the fluid. After [117] the value of C_D is approximately one for Re in the range of 10^3 to $6 \cdot 10^3$ and 1.2 for Re in the range of about 10^4 to $2 \cdot 10^5$ (smooth cylinder subjected to constant, uniform, free stream flow).

For a final fatigue analysis of the structure, the coefficient C_D has to be chosen as $C_D = 1.25$ according to estimates of the U.K. Department for Energy Guidelines [144].

For OWECS fatigue analysis, done by Kühn [86], a drag coefficient $C_D = 0.7$ was chosen, whereas for extreme condition and a tower under SWL of smooth surface without marine growth a coefficient $C_D = 0.6$ was applied.

If hydrodynamic damping should be included, the expression for q_D (6.50) has to be modified:

$$q_{DH} = C_D \rho_f r |\dot{x}_r| \dot{x}_r , \tag{6.52}$$

where $\dot{x}_r$ is the relative velocity of fluid and structure:

$$\dot{x}_r = \dot{x}_f - \dot{x}_s . \tag{6.53}$$

Here, $\dot{x}_s$ indicates the velocity in x_1-direction of the structure.

Flow superposition (Morison's formula)

According to Morison et al. [101] for a cylinder in a plane flow field with the free stream velocity $\dot{x}_f = \dot{x}_f(t)$, the total time-varying load per unit length on a cylinder may be expressed by superimposing the inertial and drag loadings, expressed by (6.48) and (6.50):

$$q = q_D + q_M = \underbrace{C_D \rho_f r |\dot{x}_f| \dot{x}_f}_{\text{drag force on fixed structure}} + \underbrace{C_M \rho_f \pi r^2 \ddot{x}_f}_{\text{inertia force on fixed structure}} . \qquad (6.54)$$

For a more realistic representation, the inertia force is composed of two parts: *hydrodynamic mass* of the member in motion and *inertia force* on a stationary member in an accelerating fluid. The hydrodynamic or added mass represents the force generated by the moving member in surrounding water. The added mass has the effect of increasing the actual mass of the member and can be represented as a force in an opposite direction to the motion of the member. The inertia force due to the fluid acceleration arises from the pressure gradient in the accelerating fluid which causes distortion of the stream lines around the stationary member. Any other forces caused by the wave motion such as lift forces, buoyancy forces, forces due to currents, interaction forces between members and boundaries will not be taken into account:

$$q = q_{DH} + q_M + q_S$$
$$= C_D \rho_f r |\dot{x}_r| \dot{x}_r + C_M \rho_f \pi r^2 \ddot{x}_f + (C_M - 1) \rho_f \pi r^2 \ddot{x}_s . \qquad (6.55)$$

This is *Morison's formula* with structural motion, hydrodynamic added mass and hydrodynamic damping. This equation is valid for most bottom-mounted OWEC support structures loaded by higher waves.

For structure members for which the ratio of significant length (the diameter in case of a cylinder) to the wavelength is small the structure can be considered as a hydrodynamically transparent, i.e. the structure movement causes no modification in the shape of the wave. In this case the Morison formula is used to determine the wave forces on the structure and the range for which it is valid is defined as:

$$\frac{D}{\lambda} < 0.2 , \qquad (6.56)$$

where D is the significant length of the interaction of the SWL with the structure (for OWECS usually the diameter of the tower below SWL) and λ is the wavelength. For large diameter structures with ratio $\frac{D}{\lambda} > 0.2$ (such as caissons of gravity support structures and tower structures with large diameter in small waves) inertia forces are much more dominant than drag forces and the changes in shape of the waves due to diffraction effects have to be taken into account, therefore the diffraction theory has to be used to determine wave forces on the structure.

The drag force , proportional to the square of the water particle velocity $\dot{x}_f$, is non-linear even if the water particle velocity is sinusoidal according to the linear wave theory. This generates difficulties in a linear frequency domain and probabilistic analysis.

Therefore, a linearisation procedure can be applied to obtain the linearised Morison force. If a linearised drag coefficient C_D^l exists the Morison force can be expressed as:

$$q = q_M + q_s + C_D^l \dot{x}_r + \underbrace{C_D |\dot{x}_r| \dot{x}_r - C_D^l \dot{x}_r}_{res} , \qquad (6.57)$$

where *res* can be interpreted as a residual function which has to be minimised. This minimisation can be established, e.g., by a least-square approach:

$$\sum_{i=1}^{N} (res_i)^2 = \sum_{i=1}^{N} \left(C_D |\dot{x}_{ir}| \dot{x}_{ir} - C_D^l \dot{x}_{ir} \right)^2 \rightarrow min \qquad (6.58)$$

$$\Rightarrow \langle res_{,C_D^l}^2 \rangle = 2 \langle C_D^l \dot{x}_r^2 - C_D \dot{x}_r^2 |\dot{x}_r| \rangle = 0 , \qquad (6.59)$$

with $\langle y \rangle$ denotes the time average of a function $y = y(t)$. Thus, the linearised drag coefficient follows:

$$C_D^l = C_D \frac{\langle \dot{x}_r^2 |\dot{x}_r| \rangle}{\langle \dot{x}_r^2 \rangle} . \qquad (6.60)$$

Assuming a Gaussian distribution with zero mean, the time average of $\dot{x}_r^2$ and $\dot{x}_r^2 |\dot{x}_r|$ can be expressed as:

$$\langle \dot{x}_r^2 \rangle = \sigma_r^2 \quad \text{and} \qquad (6.61)$$

$$\langle \dot{x}_r^2 |\dot{x}_r| \rangle = \sqrt{\frac{8}{\pi}} \sigma_r^3 , \qquad (6.62)$$

with $\sigma_r = \sigma(\dot{x}_r)$ is the standard deviation of the relative velocity of the fluid $\dot{x}_r$ and hence:

$$C_D^l = C_D \sqrt{\frac{8}{\pi}} \sigma_r . \qquad (6.63)$$

To obtain the linearised drag coefficient C_D^l the distribution of the relative velocity $\dot{x}_f$ should be known. Here is assumed that the water elevation function η has Gaussian distribution and hence the water particle velocity $\dot{x}_f$ as well. However, the relative velocity $\dot{x}_r$ may not have Gaussian distribution and therefore the standard deviation $\sigma(\dot{x}_r)$ has to be obtained with an iterative procedure. The solution procedure can be started with the standard deviation of the water particle velocity $\sigma(\dot{x}_f)$ and iterate until convergence of σ_r is obtained.

The linearised Morison equation can be written as:

$$q = C_D \rho_f r \sqrt{\frac{8}{\pi}} \sigma_r \dot{x}_r + C_M \rho_f \pi r^2 \ddot{x}_f + (C_M - 1) \rho_f \pi r^2 \ddot{x}_s . \qquad (6.64)$$

6.3.4 Correlation of wind and wave loading

Since sea waves are caused by a wind blowing a sufficiently long time, the sea states are related to the wind parameters and correlated wave and wind loading conditions can be defined on the structure. But wind and wave *directions* are not correlated, because the wave direction strongly depends on the coast line, while the wind direction may not. For OWCS simulations in this work, the worst case where the wind direction coincides with the wave direction is assumed.

Significant wave height H_s and zero crossing period T_0 which define the sea state can be determined from wind characteristics such as mean wind speed, fetch and duration. The relation between the sea state parameters H_s and T_0 and the mean wind speed at $19.5m$ height is presented as empirical formulas derived by Neumann and Pierson [102] (6.36) and (6.37). These relationships allow the sea state to be correlated with the mean wind speed and hence appropriate combination of wind and wave loading can be defined for the numerical model.

6.3.5 Sea ice loading

To analyse the ice impact hazard, it is necessary to know ice speed, size, and material properties. Drifting ice travels at speeds from 1% to 7% of the wind speed [144]. With the usual concentration of Na_2SO_4 sea ice has a compressive or rupture strength from $1.4MPa$ to $2.8MPa$, but this varies with salt concentration and the rate of impact loading [110]. The American Petroleum Institute [5] has recommended the following formula for calculating the horizontal force F_h on structures subjected to the impact of ice:

$$F_h = C_i \sigma_{ci} A_0 , \tag{6.65}$$

with C_i is a coefficient in the range of 0.3 to 0.7 which accounts for loading rate; σ_{ci} is the compressive strength or rupture stress for the ice and A_0 is the area of the structure exposed to the impacting ice. The gravity foundation of concrete tolerates impacting ice. This leg geometry leads to effective stress rupturing of the impacting ice with minimal damage to the structure, see Fig. 6.13.

Besides the impact of floating ice, there are other ice hazards to a structure in the Baltic or North Sea. For instance, the uplift force on the structure due to the buoyancy of accumulated ice attached to the tower around the water line (the specific gravity of sea ice ranges from 0.89 to 0.92 [144]). Although this uplift force may be offset by added gravity loads on the superstructure due to ice accumulation there, this accretion also increases wind loads because of the increase of exposed area. Further, ice accumulation on the tower or foundation increases wave and current loads for the same reason.

Ice also causes abrasion in various forms. Cyclic freezing and thawing leads to cracking and spalling of concrete structures, due to the expansion of freezing water in cracks, pores, or capillary cavities. Some of this entrained water is the excess required for hydration of the cement and may be minimised by careful choice of the ingredients for the concrete.

The ice thickness in the Baltic Sea (at the German coast) varies from 0 to $0.7m$ [31], whereas in the North Sea, the ice thickness is in the range of 0 to $0.4m$ [84]. Therefore, ice loading should not be neglected in some site areas of the North Sea.

6.4 Subsoil in the North Sea

The subsoil which is relevant for the foundation of OWECS in the North Sea consists of Holocene and Pleistocene beds. While the Holocene sediments are usually not sustainable, the Pleistocene is sustainable, but with widely spreading properties. Due to changing sedimentation conditions over the last 50000 years, and filled sea channels, the load capacity of the subsoil may change locally. Therefore, a soil exposure for the particular location is necessary.

For the first phase of planning the foundation, Sindowski [121] made many soil exposures in the North Sea, and presents four associated geological sections. For soil characteristic values of the North Sea, refer to the tables of [129] or [68].

Holocene

The Holocene bed is the top layer of the subsoil, with a variant thickness of approximately $1m$ - $15m$. Soil types of the Holocene are mainly sands, clayed/poor clayed sediments, in some areas also gravel, particular clam shell. Sediments are often loose to middle dense bedded, and of varying thickness. As mentioned above, the Holocene layer is normally not sustainable.

Pleistocene

The next layer (under the Holocene) is called Pleistocene bed with a thickness of up to $70m$. Types of soil of the Pleistocene bed are sands, boulder clay, so-called Lauenburger clay and thin layers of peat. The cohesive soils are generally over consolidated and of good loading capacity. Filled channels in the Pleistocene bed are often of poor loading capacity, and some erratic blocks can be a serious problem for the construction of driven pile foundations.

6.5 Structure

The structure of offshore wind turbines consists of two main substructure systems: rotor and tower system. The rotor system represents usually three rotor blade elements and the nacelle, while the tower system includes tower and foundation, see Fig. 6.9.

6.5.1 Foundation

A substantial part of the construction costs is generally assigned to the foundation. According to experience with conventional buildings on the mainland these are 10% to 30% of the carcass costs, depending upon building ground condition [143]. The portion of carcass costs in the main sea can be even higher due to the special boundary conditions. Thus a high economic and technical importance is attached to the investigation of the ground characteristics and the correct selection of the type of

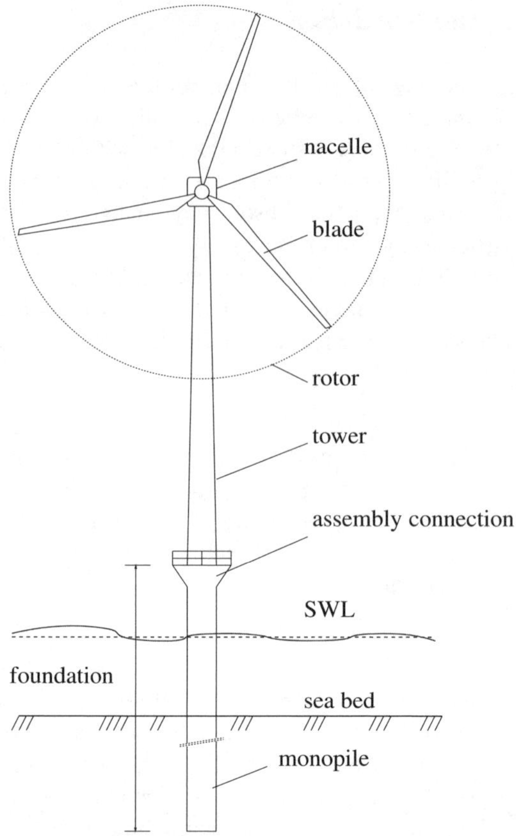

Fig. 6.9. Components of an OWECS

foundation. For an economic realisation of offshore wind parks it is advisable that these kinds of foundations are to be developed, which can be adapted locally to the respective foundation-relevant boundary conditions despite a high-level prefabrication. This requires sufficient knowledge of the sea bed characteristics and their conversion in all planning phases in the context of foundation planning.

The foundation transfers the loads of the OWECS into the building ground and is connected by an assembly connection with the tower, see Fig. 6.9. The tower of an OWECS can be carried out as monotower or lattice tower. A braced tower construction has a lattice wing unit or laterally supports at the tower, spread at the lower end, which enlarges the stiffness of the load carrying system.

Monopile

The monopile foundation is a simple construction: it consists of a steel pile with a diameter of between $3.5m$ and $4.5m$. This pile is driven approximately $3.5D$ to $8D$

(with D is the diameter of the monopile) into the sea bed depending on the type of underground, see Tab. 6.3. A monopile foundation effectively extends the turbine tower under water and into the sea bed, see Fig. 6.10. An important advantage of this foundation is that no preparations of the sea bed are necessary. On the other hand, it requires heavy duty piling equipment, and the foundation type is not suitable for locations with many large boulders in the sea bed. If a large boulder is encountered during piling, it is possible to drill down to the boulder and blast it with explosives.

type of soil	clamping length
average	$6D$
stiff clay	$3.5D$ - $4.5D$
soft silt	$7D$ - $8D$

Table 6.3. Clamping lengths for monopile foundation for different soil conditions

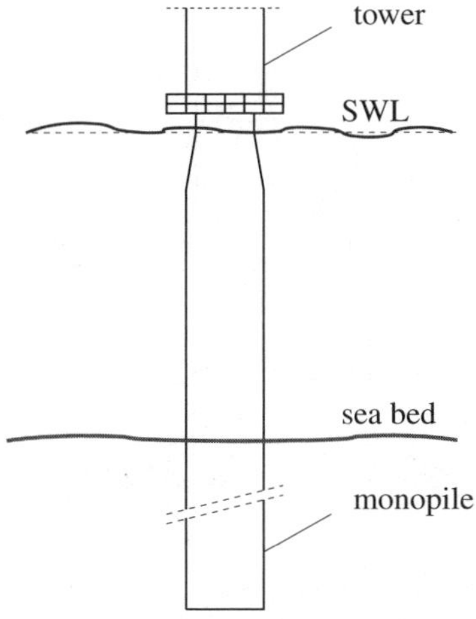

Fig. 6.10. Schematic sketch of an OWECS monopile foundation

The dimensioning factor of the foundation varies from the North Sea to the Baltic Sea. In the North Sea it is the wave size that determines the dimension of the monopile. In the Baltic Sea the pack ice pressure decides the size of the foundation. This is the reason why monopile foundation costs increase more rapidly in dependency of the water depth in the Baltic Sea than in the North Sea. In contrast to other types of foundation, erosion will normally not be a problem with this type of foundation.

A $2.5MW$ pilot project with five Danish wind turbines using the monopile technology has been installed in the Baltic sea south of the Swedish island of Gotland

[42]. Using the mono pile foundation technique at Gotland involved drilling a hole of $8m$ to $10m$ depth for each of the turbines (Wind World $500kW$). Each steel pile is slotted into the solid rock. When the foundations are in place the turbines can be bolted on top of the monopiles [42].

Tripod

The tripod foundation draws on the experiences with light weight and cost efficient three-legged steel jackets for marginal offshore fields in the oil industry. A steel frame is connected to the turbine tower, and transfers the forces from the tower into three steel piles, see Fig. 6.11. The three piles are driven $10m$ to $20m$ into the sea bed depending on soil conditions and ice loads.

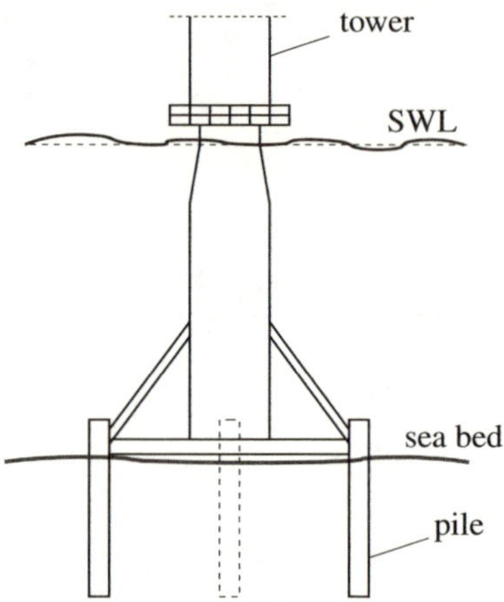

Fig. 6.11. Schematic sketch of an OWECS tripod foundation

The advantage of the three-legged model is that it is suitable for larger water depths. Additionally, only a minimum of preparation is required at the site before installation. For a so-called multipile technology, the foundation is anchored into the sea bed using a relatively small steel pile ($0.9m$ diameter) in each corner, see Fig. 6.11. Because of the piling requirement, the tripod foundation is not suited for locations with many large boulders. Erosion will normally not be a problem with this type of foundation, but it is not suitable at water depths lower than $6m$ to $7m$. The main reason for this is that service vessels at low water depths will face problems approaching the foundation due to the steel frame.

Basic difference between costs in the North Sea and the Baltic Sea is that waves determine dimensioning in the North Sea, whereas ice is decisive in the Baltic Sea.

Gravitation foundation of steel

Most of the existing offshore wind parks use gravity foundations. A new technology offers a similar method to that of the concrete gravity caisson. Instead of reinforced concrete it uses a cylindrical steel tube placed on a flat steel box on the sea bed, see Fig. 6.12.

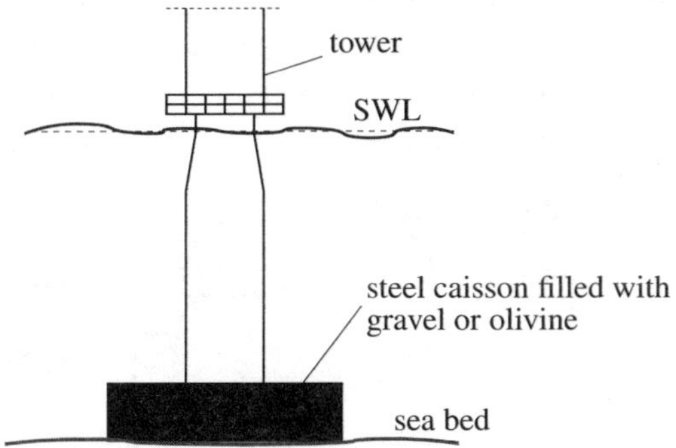

Fig. 6.12. Schematic sketch of an OWECS steel caisson gravity foundation

A steel gravity foundation is considerably lighter than concrete foundations. Although the finished foundation has to have a mass of around $10^6 kg$, the steel structure will only have a mass of $8 \cdot 10^4 kg$ to $10^5 kg$ for water depths between $4m$ and $10m$. (Another $10^4 kg$ have to be added for structures in the Baltic Sea, which require pack ice protection). Relatively low weight allows barges to transport and install many foundations rapidly, using the same fairly lightweight crane used for the erection of the turbines. Gravity foundations are filled with olivine, a very dense mineral, which gives the foundations sufficient weight to withstand waves and ice pressure. The base of a foundation of this type will be $14m$ by $14m$ (or a diameter of $15m$ for a circular base) for water depths from $4m$ to $10m$ (calculation based on a wind turbine with a rotor diameter of $65m$).

Advantage of the steel caisson solution is that the foundation can be made onshore, and may be used on all types of sea bed although sea bed preparations are required. Silt has to be removed and a smooth horizontal bed of shingles has to be prepared by divers before the foundation can be placed on the site.

The sea bed around the base of the foundation will normally have to be protected against erosion by placing boulders or rocks around the edges of the base. This is, of course, also the case for the concrete version of the gravitation foundation; it makes the foundation type relatively costlier in areas with significant erosion. Dimensioning factor which decides the required strength and weight of the foundation is not the turbine itself but ice and wave pressure forces.

Gravitation foundation of concrete

The first offshore pilot projects used concrete gravity caisson foundations. The Vindeby offshore wind farm and the Tunoe Knob wind farm are examples of this traditional foundation technique. Caisson foundations were built in a dry dock near the sites using reinforced concrete and were floated to their final destination before being filled with sand and gravel to achieve the necessary weight. The principle is thus much like that of traditional bridge building. Foundations used at these two sites are conical to act as breakers for pack ice. This is necessary because solid ice is regularly observed in the Baltic Sea.

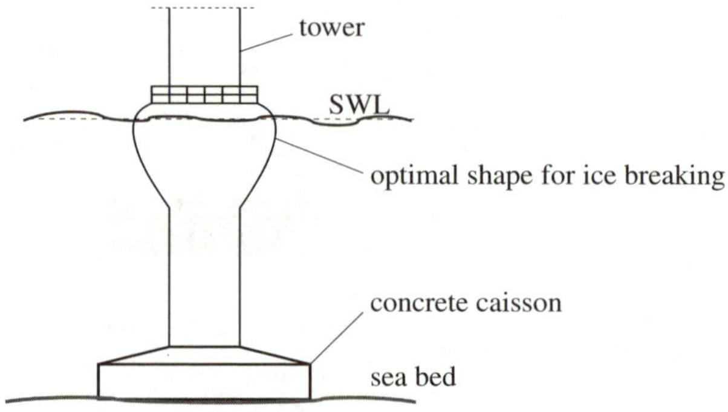

Fig. 6.13. Schematic sketch of an OWECS concrete gravity foundation

Using traditional concrete foundation techniques the cost of the completed foundation is approximately proportional with the water depth squared. The water depths at Vindeby and Tunoe Knob vary from $2.5m$ to $7.5m$. This implies that each concrete foundation has an average weight of approximately $10^6 kg$. Thus, the concrete platforms tend to become prohibitively heavy and expensive to install at water depths above $10m$.

6.5.2 Tower

The tower transfers the load from the nacelle to the foundation. For OWECS, the support structure comprises the tower and foundation. Towers for large wind turbines may be either tubular steel towers, lattice towers, or concrete towers. Guyed tubular towers are only used for small wind turbines (battery chargers etc.).

Most large onshore wind turbines and most offshore wind turbines are delivered with *tubular steel towers* , which are manufactured in sections of 20-30 metres with flanges at either end, and bolted together on the site. These towers are conical, i.e., with their diameter increasing towards the base.

Lattice towers are manufactured using welded steel profiles. Basic advantage of lattice towers is cost, since a lattice tower requires only half as much material as a freely standing tubular tower with a similar stiffness. Basic disadvantage of lattice towers in the offshore environment is the increased ice loading, and, furthermore, the high costs for anti-corrosion sealing.

The optimum height of the tower is a function of tower costs per meter and how much the wind locally varies with the height above ground level, i.e. the average local terrain roughness (large roughness makes it more useful with a taller tower). Rotor blades on turbines with relatively short towers will be subject to very different wind speeds (and thus different bending) when a rotor blade is in its top and in its bottom position, which will increase the fatigue loads on the turbine.

6.5.3 Nacelle and wind turbine rotor

The nacelle contains the key components of the wind turbine, including the gear-box, and the electrical generator. Service personnel may enter the nacelle from the tower of the turbine or form a helicopter deck located on the nacelle.

The wind turbine rotor consists of the rotor blades and the hub. The rotor blades capture the wind and transfer its power to the rotor hub. On a modern $1MW$ wind turbine each rotor blade measures about $27m$ in length.

6.6 Numerical example

The reference case for the numerical investigations in this subsection is a $3MW$, three-bladed wind turbine. For the numerical example, an OWECS with a height above SWL of $55m$ in $25m$ deep water is chosen (see Fig. 6.17). The OWECS has a monopile foundation and a slender steel tower. Thus, the structure characteristic is soft/soft. A three blade rotor with a rotor diameter of $80m$ is considered.

The monopile foundation has a diameter of $3.5m$, while the diameter of the tower is in a range of $3.5m$ (wall-thickness $75mm$) at the bottom to $2.8m$ (wall-thickness $20mm$ at the top, see Tab. 6.6. The construction is similar to the WTS80M wind turbine, as described in Kühn [86].

6.6.1 Offshore environment

In the offshore environment the wind turbines interact with the surrounding environment, which are such as wind and wave loadings or ice loading, especially in the Baltic Sea. Thus, analysis and design calculations of OWECS are based on state of the art wind energy and offshore technology, where multi-related models are used. But not only special loading conditions have to be considered, also the different sub-soil is of particular importance.

Wind

For the reference site, the mean wind speed at $19.5m$ height is equal to $21.2m/s$ according to the sea state (see next paragraph). The mean wind speed at hub height

($55m$) can be calculated with (6.1), (6.2) and (6.44), where $k \approx 0.4$ and $z_0 = 0.2mm$ for the offshore environment. The resulting mean wind speed at hub height is $\dot{x}_{55} = 23.1m/s$.

Before simulating the dynamic behaviour of the rotor system, the wind field loading (turbulent wind field) has to be determined. Here, the spatial variation of the turbulence intensity is taken into account, see Sec. 6.2.2. Thus, the simulation of a three-dimensional wind field is necessary. The computation of the wind field follows Shinozuka [118], where time-series of the wind speed on several points are generated. These points lay on a plane, perpendicular to the plane of main wind speed, and according to the frozen turbulence theory of Taylor [130], they travel with the mean wind speed. In order to calculate the needed time-series, the average wind speed as a function of height over SWL must be given, and additionally, the time- and space-dependent variation of the wind speed. The discrete time-series for a chosen number of points on the rotor plane are generated with an inverse fast Fourier transformation, see Meyer [99].

When using polar coordinates (which is favourably for this type of problem), a cylinder – where the cylinder axis represents the axis of time – is filled with a grid of wind speeds. If then wind speed data of a specific point in time and space is needed for the OWECS simulation, the turbulent wind speed is calculated via interpolation of neighbouring data points.

The reaction forces (bending moment M_x and shear force F_x) from the wind loading on the rotor blades are applied as external time dependent forces at the tower top, see Fig. 6.16.

Waves

The model sea state is chosen according to the scatter diagram of United Kingdom Offshore Operators' Association, see Patel [109], with a significant wave height $H_s = 9m$, a zero-crossing period $T_0 = 11s$.

The wave loading is determined by the Morison formula for relatively moving elements with the assumption of linear wave theory, see Secs. 6.3.1 and 6.3.3. In order to apply the Morison formula, the structure is assumed to be hydrodynamically transparent which is the case when the tower diameter is less than one fifth of the wave length. The distributed wave loading is supposed to act in x_1 direction and according to the linear wave theory it is applied up to the mean sea level (SWL).

The random function of the wave elevation is generated from the Pierson-Moskowitz wave spectrum corresponding to the sea state, see Sec. 6.3.2. Water kinematics is determined from the water elevation by linear wave theory. Water velocities and accelerations are then used to compute the Morison force on a relatively moving member. The structural response is obtained by time integration of the dynamical system of equation using the HHT-α-method , see Sec. 1.4.2. As simulation results, the displacement time history at tower top is plotted as a reference. As a matter of course, at all structural points of the wind turbine and the soil of the near-field, the displacement and stresses can be calculated. For example, if a fatigue analysis of the structure should be done, the (maximum) stress at tower base can be taken into account.

The wave loading is calculated using the linear wave theory assuming that small but frequently occurring waves form the major part of the loading. The water elevation which is considered a random function is generated from the Pierson-Moskowitz spectral density function for fully developed sea.

The random function of water elevation is generated from the Pierson-Moskowitz spectrum by superposition of harmonic waves with amplitudes equal to the Fourier coefficients with equations (6.46) and (6.47). Criteria for determining the number N of superposed harmonic waves in these equations is the integrated area of the spectrum to be at least 95% of the whole spectral area.

Wave numbers k_i corresponding to the frequencies ω_i of the harmonic waves are calculated form the non-linear dispersion relation (6.31), using the Newton method. The non-linear relation is used, because the water depth is less than half the wave length (shallow water). Water particle velocities and accelerations are determined from the generated function of wave elevation by the linear wave theory (6.21) and (6.23) for shallow water.

parameter	symbol	value
inertia coefficient	C_M	1.93
drag coefficient	C_D	1.25
wet diameter	D	$3.5m$
cross sectional area	A	$9.62m^2$
fluid mass density	ρ_f	$1000\,kg/m^3$

Table 6.4. Parameters for Morison force

The wave loading on the structure is defined by the Morison force, see Sec. 6.3.3; parameters, necessary to determine the Morison force are listed in Tab. 6.4. Finally, the calculated concentrated Morison loads at SWL (indicated with Q_1 in Fig. 6.17) and in $25m$ water depth (indicated with Q_8) are plotted in Fig. 6.14.

Soil

For the reference site, a $5m$ thick layer of saturated loose sand over bolder clay is assumed, see Fig. 6.17. The material parameters of the soil can be found in Tab. 6.5. The Young's modulus of soil can be calculated via[4]

$$E = \frac{1 - \nu - 2\nu^2}{1 - \nu} E_s \,, \tag{6.66}$$

where E_s is the given stiffness modulus ($E_s \approx 50MPa$ for loose sand, and $E_s \approx 100MPa$ for bolder clay). These values are chosen for a representative numerical study. For realistic values, a ground survey has to be done at the site of erection.

---

[4] see, e.g.,[113, 112]

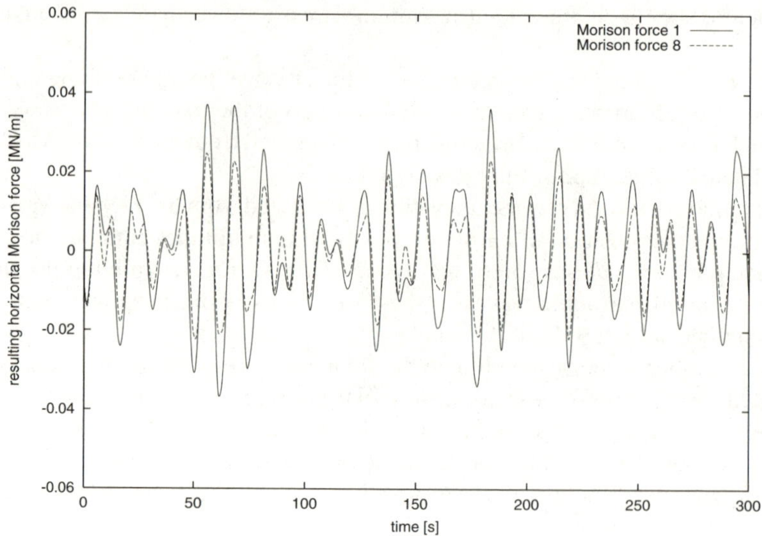

Fig. 6.14. Concentrated Morison loads Q_1 and Q_8

material	density ρ [kg/m^3]	Young's modulus E [MPa]	Poisson's ratio ν [$-$]	
steel	7850	210000	0.3	construction
loose sand	1800	37.15	0.48	saturated
boulder clay	2200	74,30	0.48	soil

Table 6.5. Material parameters

6.6.2 Discretisation

The discretisation and simulation of the rotor system dynamics follow Meyer [99]. For a simulation time of $300s$, the resulting horizontal load F_{x_1} is displayed in Fig. 6.15. Additionally, the horizontal load F_{x_2}, vertical load F_{x_3} and bending moments M_{x_1} and M_{x_2} are applied at the tower top as time-dependent load, see Fig. 6.16.

The tower of the wind turbine is modelled with three-dimensional two node Euler-Bernoulli beam elements . The used beam element is a standard finite element, where the stiffness- and mass matrix are described in many textbooks [153, 13], therefore they are not repeated here. Element numbering and element lengths are shown in Fig. 6.17. For the welded steel chimney tower with varying diameter and wall thickness, the element geometries are collected in Tab. 6.6. The $20m$ long monopile foundation is discretised with the same beam elements, see Fig. 6.17.

The beam elements of the foundation are coupled with three-dimensional solid brick elements, see Fig. 6.16, which form the near-field. This eight-node element is based on an isoparametric formulation using linear shape functions. The element

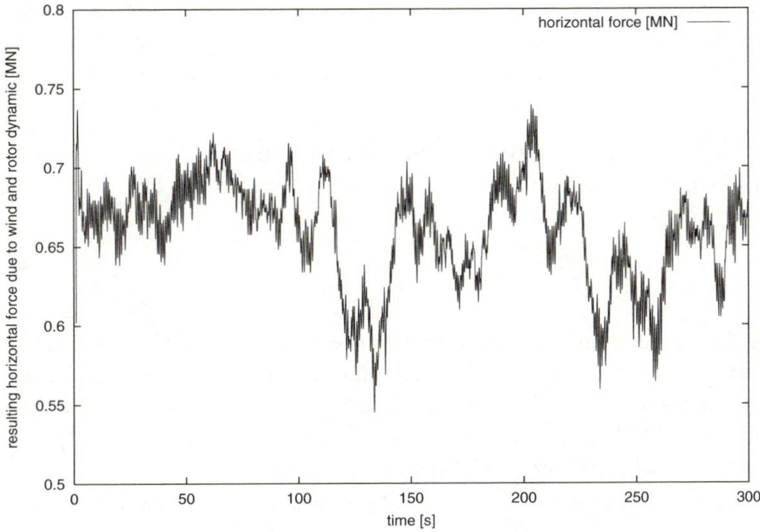

Fig. 6.15. Horizontal load F_{x_1} due to wind and rotor dynamics

incorporates all three translation degrees of freedom at each node, for a total of 24 local degrees of freedom. The 24×24 element stiffness matrix is evaluated using a fourth-order accurate $2 \times 2 \times 2$ Gaussian quadrature rule. This formulation results in stress calculations at eight integration points. For this three-dimensional discretised near-field, a non-linear material behaviour could be assumed.

The outer three-dimensional brick elements of the near-field are coupled with two dimensional four node scaled boundary finite elements (far-field), see Figs. 6.18, 6.19 and 6.20, as described in Sec. 3.6. Thus, no reflection of waves occurs at the artificial boundary (near-field/far-field interface). The shape of the top soil layer (loose sand) is dependent on the distance of the near-field/far-field interface from the scaling centre (see Figs. 6.18 and 6.19), but for the chosen discretisation and geometry, the different shapes have no significant influence on the results, see next subsection.

To choose the time-step for SBFE analysis, the smallest distance $\min d_e$ of SBFEs to the scaling centre is the limitation, and $\Delta t \leq \frac{\min d_e}{15c}$, where c is the pressure wave speed for elastodynamic calculations, given by

$$c_p = \sqrt{\frac{\lambda + 2\mu}{\rho}} \text{ , with} \tag{6.67}$$

$$\lambda = \frac{E\nu}{(1+\nu)(1-2\nu)} \text{ and} \tag{6.68}$$

$$\mu = \frac{E}{2(1+\nu)} \ . \tag{6.69}$$

With material parameters, given is Tab. 6.5, the resulting pressure wave speeds are: $c_{p_s} = 425.8\frac{m}{s}$ (loose sand) and $c_{p_c} = 544.6\frac{m}{s}$ (boulder clay). With $\min d_{e_s} = 10m$ for loose sand (see Fig. 6.18) and $\min d_{e_c} = 11.2m$ for boulder clay, the resulting minimal time-step is $\Delta t \leq 1.4 \cdot 10^{-3}s$ $\left(= \frac{\min d_{e_c}}{15 \cdot c_{pc}}\right)$.

For the FE part of the OWCS simulation, a time-step $\Delta t \leq \frac{\min \ell_e}{10 \cdot c_b}$ is adequate, where c_b is the propagation speed of bending waves in the tower structure. This speed is given by

$$c_b = \sqrt{\omega^4 \frac{b}{\rho h}} \, , \tag{6.70}$$

where ω is the maximum frequency introduced by outer forces. The maximal frequency which can be incorporated in the numerical model depends on the resolution of measured forces which act on the structure. For the presented simulation, the resolution is $\frac{1}{10}s$, which results in a maximal frequency $\omega_{max} = 31.4\frac{rad}{s}$. In (6.70) B stands for bending stiffness of the structure, and is given by Eq. 5.13., ρ is the density and h the thickness of considered configuration. The resulting time-step is $\Delta t \leq \frac{2.5m}{10 \cdot 31.36\frac{m}{s}} = 8 \cdot 10^{-3}s$. Hence, due to SBFE limitations, for all simulations a time-step size of $\Delta t = 1 \cdot 10^{-3}s$ is chosen.

element-number	element-length	upper diameter	wall-thickness	position
[−]	[m]	[m]	[mm]	
1	4.50	3.50	30	
2	6.00	2.80	20	above
3	12.00	2.80	25	
4	12.00	2.80	30	SWL
5	12.00	2.80	60	
6	8.50	2.80	60	
7	2.50	3.24	60	
8	2.50	3.37	60	under
9	2.50	3.50	75	
10	2.50	3.50	75	SWL
11	5.00	3.50	75	
12	4.00	3.50	75	
13	6.00	3.50	75	
14	5.00	3.50	75	
15	7.50	3.50	75	soil
16	7.50	3.50	75	

Table 6.6. Element geometries of the tower (element numbering see Fig. 6.17)

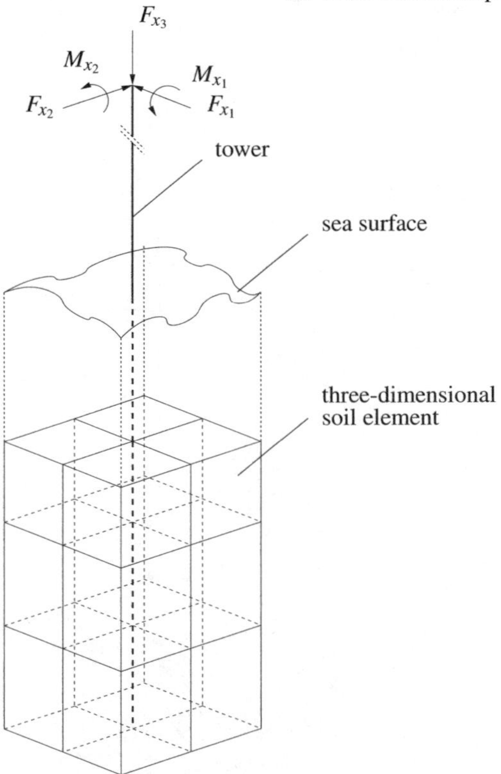

Fig. 6.16. Part of the soil near-field: three-dimensional discretisation

6.6.3 Results

Fig. 6.18 depicts a discretisation of the soil for a near-field width of $20m$ (indicated with discretisation A). Here, the scaling centre has a horizontal distance of $10m$ to the near-field/far-field interface. A discretisation with $30m$ near-field width is illustrated in Fig. 6.19. Fig. 6.21 shows the results (calculated horizontal displacement at tower top) for both discretisations. No significant difference between discretisation A and B can be observed, therefore the width of the near-field of discretisation A is sufficient.

To examine the influence of the finite element mesh of the near-field solid elements, another simulation with a refined mesh (Fig. 6.20) is performed. The results (Fig. 6.22) show again no significant deviation to the coarse mesh (discretisation A).

The recursive algorithm of section 3.7.1 in conjunction with sparse influence matrices (see Sec. 3.7.2, zero-element threshold $\varepsilon_z = 1 \cdot 10^{-4}$) is used. The entries of the influence matrices grow linear from a certain time t_m. This time-station t_m is determined as $t_m = 0.047s$, which corresponds to a linear behaviour after $m = 188$ time-steps. With a total of $3 \cdot 10^5$ time-steps (simulation time $t_n = 3 \cdot 10^5 \cdot 1 \cdot 10^{-3}s = 300s$) to be computed, the reduction of computational effort – according to (3.136) and (3.137) – exceeds 99%.

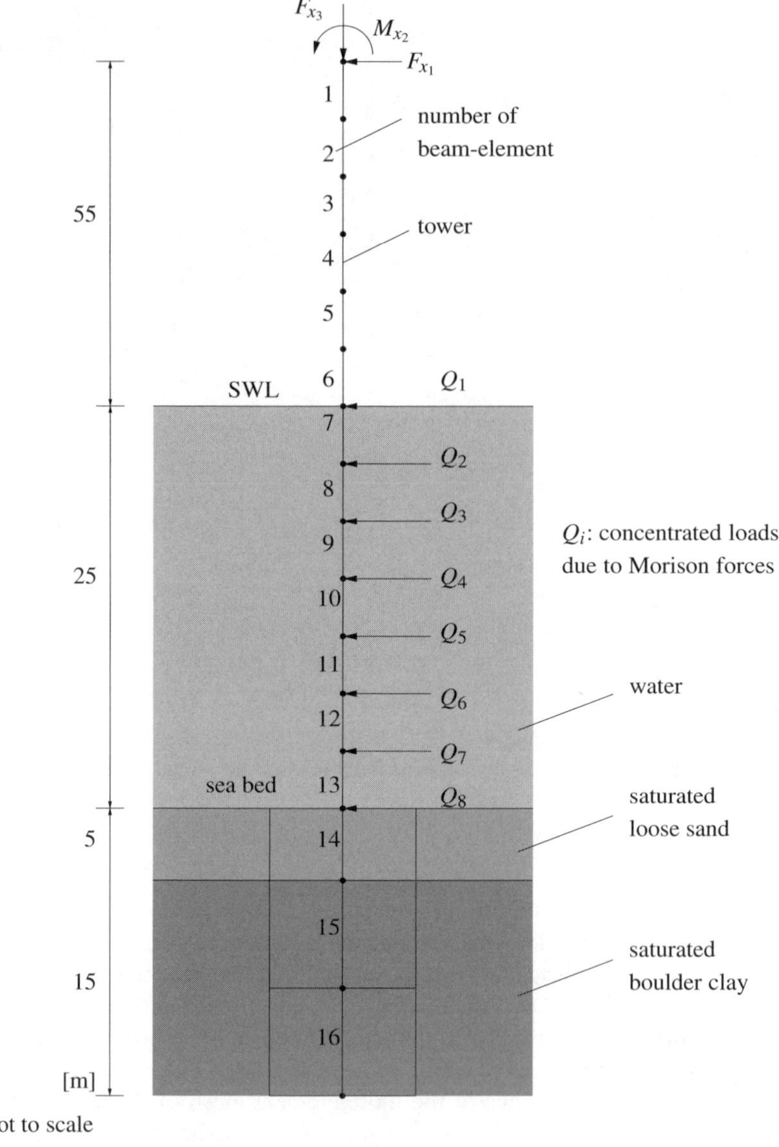

Fig. 6.17. Schematic sketch of soil-tower system discretisation (longitudinal section)

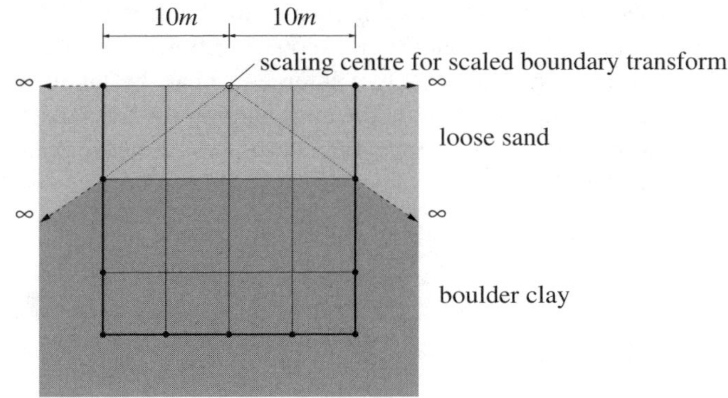

Fig. 6.18. Discretisation A of soil, longitudinal section: coarse mesh, width of near-field 20*m*

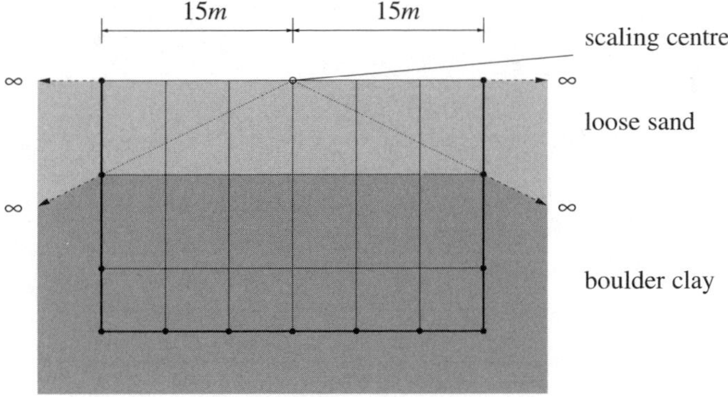

Fig. 6.19. Discretisation B of soil, longitudinal section: coarse mesh, width of near-field 30*m*

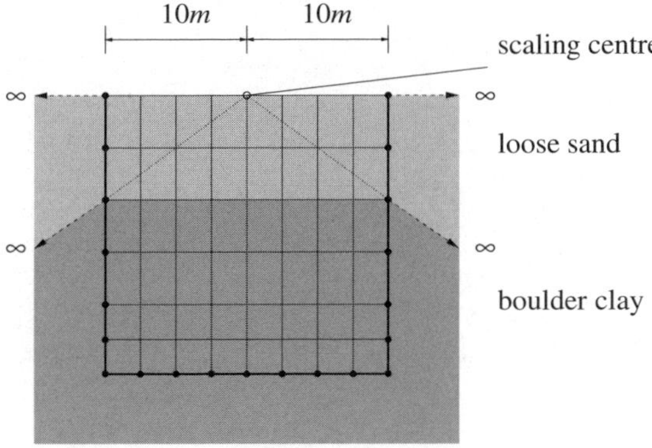

Fig. 6.20. Discretisation C of soil, longitudinal section: refined mesh, width of near-field 20*m*

Therefore, this numerical model is suitable for an OWECS simulation with realistic wind- and wave loadings and consideration of the structural dynamic behaviour. Furthermore, the soil-structure interaction can be taken into account. A postprocessed analysis may include the computation of stress distribution in the tower system, where three-dimensional effects (like asymmetric openings in the tower) could be considered.

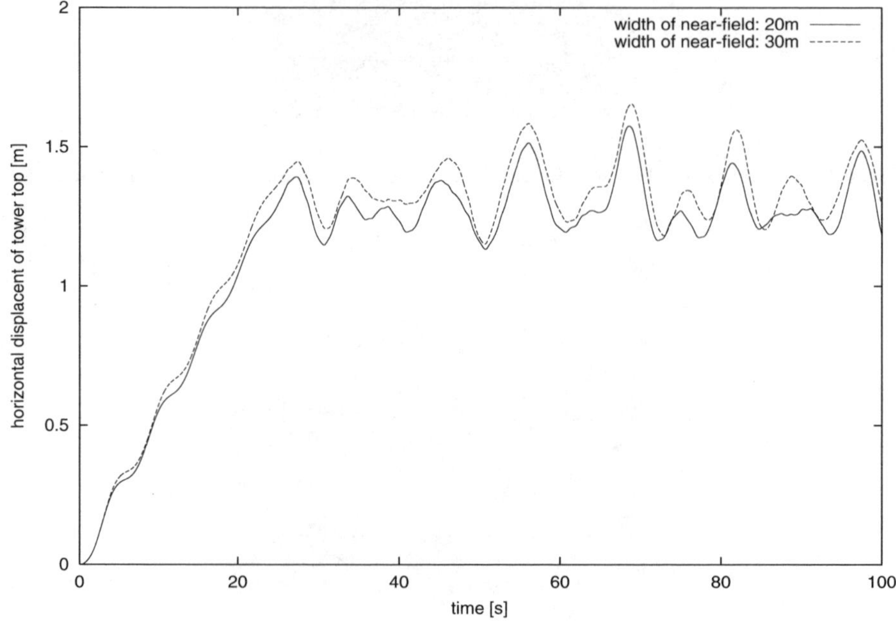

Fig. 6.21. Calculated horizontal displacement at tower top for discretisation A and B for first 100s

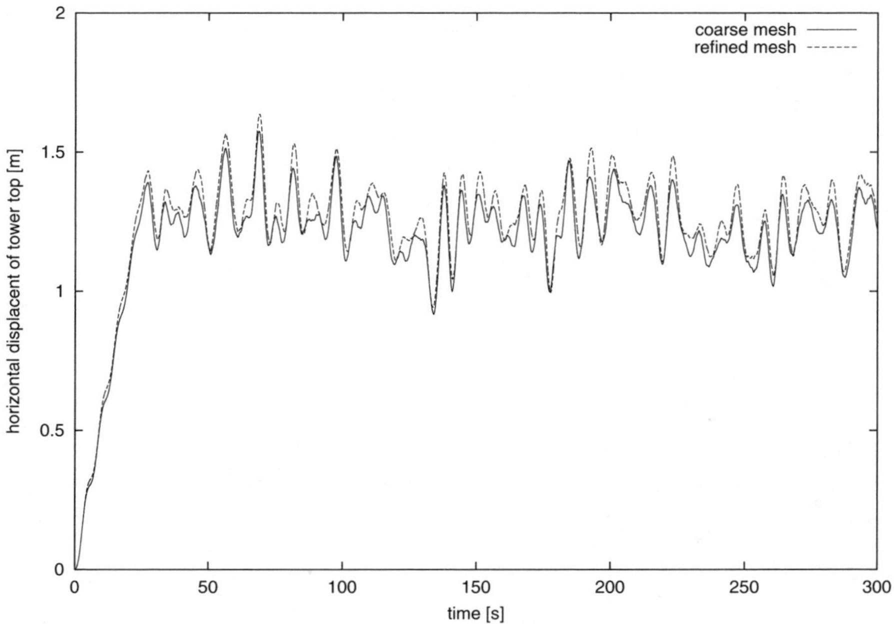

Fig. 6.22. Calculated horizontal displacement at tower top for discretisation A and C

7. Earthquake excited building

After a short introduction to seismic waves and ground response analysis, a numerical simulation of a three-dimensional soil-structure interaction problem follows. For a simple building – discretised with finite elements – data recorded during the Kobe earthquake are used as excitation.

7.1 Seismic waves

Seismic waves are divided into body and surface waves. There are two types of body waves: compressional or P waves , where the ground particles move into wave propagation direction and shear or S waves with a perpendicular particle motion. Surface waves are the result of body waves interacting with layer boundaries, e.g., the surface of the earth. Their magnitude decreases with increasing depth, they arrive after body waves, and transport roughly 75% of the seismic energy. Two important surface waves are the Love wave (L-wave) and the Rayleigh wave (R-wave). The Love wave moves ground particle perpendicular to the direction of motion, while the Rayleigh wave moves the particle in a retrograde ellipse in propagation direction.

7.2 Ground response analysis

During an earthquake, seismic waves are travelling from their initial fault through the rocky inner of the earth. Only the material of a very thin layer at the surface is soil. When waves arrive at the bedrock-soil interface, their amplitudes and propagation directions are changing significantly. The methods to evaluate these changes of wave propagation in the soil are subsumed under the concept of *ground response analysis*. Ground response models are used to identify the response of the upper soil-layers of the earth's crust.

The most encountered ground response models in practice are described in [85] and [152]. These models differ in their considered dimension of the problem and in their approaches to the material behaviour. The numerical soil models are basically divided into an equivalent linear or non-linear approach. Although the accurate description of the material behaviour is non-linear, an equivalent linear approximation is convenient if the model is not used to describe large deformations or strong ground motions, which is mostly the case for seismic considerations.

The propagation path of earthquake waves in a numerical model that discretises only the near-field of the examined structure, i.e., only the surface-near soil, is the following: The seismic waves enter the discretised region from below, run through this region and are partly reflected at the surface.

In these ground response models it is assumed that the initial motion at the truncation boundaries is known. Statements about these methods – how to apply the seismic input load to maintain the measured seismic surface response – are only made for simple assumptions, like one-dimensional, linear, homogeneous problems.

In more complicated practical problems, the applied initial motion at the bedrock-soil interface at the bottom of the soil model is usually assumed as the measured seismic record [21, 47, 119, 150, 151]. In these investigations the examined structures have relatively small extensions compared to the whole region affected by the earthquake, which is valid for the presented simulation as well. This allows to apply the simplified input, i.e., to take the motion, which was measured at the surface, as motion at the near-field/far-field interface bottom. Characteristic of the presented simulation with compact building is the small extension in both horizontal directions. Therefore, it is not necessary to take the wave propagation inside the near-field into account, which is in contrast to the simulation of infrastructure systems, like roads, railways or lifelines.

7.3 Numerical example

The modelled half-space soil medium is assumed to behave elastically in a linear way. Furthermore, dynamic soil-structure interaction is taken into account. Again, a coupled FE/SBFE methodology is used to compute the dynamic response in the time domain. To use a time domain simulation is obligatory for a further implementation of non-linear material behaviour. The seismic excitation is based on measured earthquake data gained from the Pacific Earthquake Engineering Research Center (PEER) strong motion data base [106] and can be generated at the near-field/far-field interface from free-field motion [47]. The seismic waves are considered to behave linearly.

For the presented numerical example, the recorded data of a seismic station near Kobe are used as excitation. A brief description of cause and impact of the Kobe earthquake is given below.

7.3.1 Kobe earthquake

Japan is positioned on the margin of the Eurasian Plate. The Philippine Sea Plate is subducted below the Eurasian plate, resulting in Japan having greater than average seismic and volcanic activity. Immediately south of Osaka Bay is a fault called the Median Tectonic Line, and it was a sudden movement along this fault that triggered the earthquake that hit Kobe 1995, January 17th [56]. This was the largest earthquake in Japan since 1923. The earthquake was not only powerful (magnitude $M = 6.9$), but with the epicentre near the city, resulted in massive damage to property and loss of life.

The earthquake was particularly devastating because it had a shallow focus. The resulting surface rupture had an average horizontal displacement of about $1.5m$ on the Nojima fault, which runs along the northwest shore of Awaji Island [140].

The worst effected area was in the central part of Kobe, a region about $5km$ by $20km$ alongside the main docks and port area. This area is built on soft and easily moved rocks, especially the port itself which is built on reclaimed ground. The proximity of the epicentre, and the propagation of rupture directly beneath the highly populated region, help explain the great loss of life and the high level of destruction. The strong ground motions that led to collapse of the Hanshin Express way also caused severe liquefaction damage to port and wharf facilities [135].

More than 102000 buildings were destroyed in Kobe, leaving over a fifth of the city population, some 300000 people, homeless, more than 5000 persons were killed and over 26000 injured. With all the costs added up, it makes the Kobe earthquake the most expensive natural disaster in the 20th century (about 250 billion EUR). Most of the deaths and injuries occurred when older wood-frame houses with heavy clay tile roofs collapsed. However, many of the structures in Kobe built since 1981 had been designed to strict seismic codes. Most of these buildings withstood the earthquake. In particular, newly built ductile-frame high rise buildings were generally undamaged.

Unfortunately, many of the buildings in Kobe had been built before the development of these strict seismic codes. The collapse of buildings was followed by ignition of over 300 fires within minutes of the earthquake. The fires were caused by ruptured gas lines. Response to the fires was hindered by the failure of the water supply system and the disruption of the traffic system [25].

For the numerical simulation, data recorded during the Kobe earthquake of the Fukui seismic station near Kobe are used.

7.3.2 System

The seismic excitation is based on measured earthquake data gained from data records of the PEER strong motion database [106]. Data of the Fukui (Hukui) station (seismograph station code FUK) are used, which is at 36.0533^{o} N, 136.2267^{o} E and $9.0m$ high above main sea level [137] on Honshu island, Japan. The closest distance to the fault rupture of Fukui station is $157.2km$.

In Figs. 7.1 to 7.3, the recorded vertical acceleration, velocity and displacement of Fukui seismic station during the Kobe earthquake are displayed. For horizontal acceleration, velocity and displacement (direction north/south and west/east) refer to appendix A (Figs. A.1 to A.6).

Figs. 7.4 and 7.5 depict the front- and side view of the discretised building, respectively. It consists of concrete and has the following material properties: Young's modulus $E = 30000MPa$, Poisson's ratio $v = 0.2$, and density $\rho = 2500\frac{kg}{m^3}$. The three-storey building has a cross-section of $10m \times 8m$, a height of $9m$, four windows in the front and two windows in one face side. The ground floor is less stiff compared to the first and second floor due to the lack of stiffening walls. This type of building

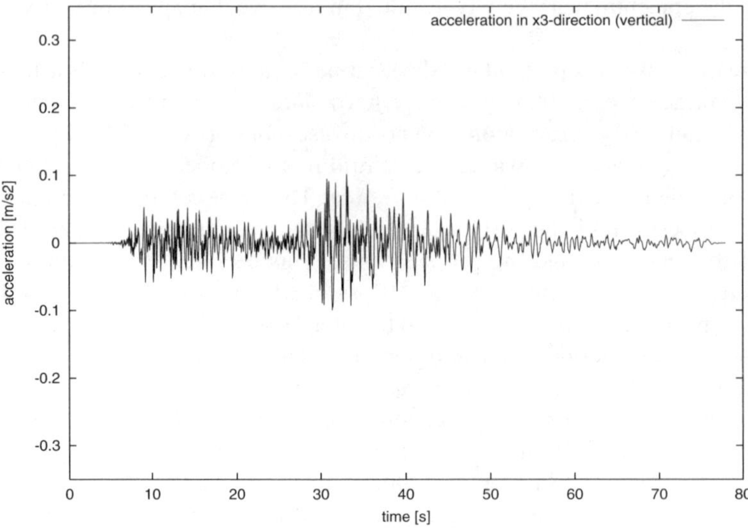

Fig. 7.1. Kobe earthquake data record (station Fukui), vertical acceleration versus time

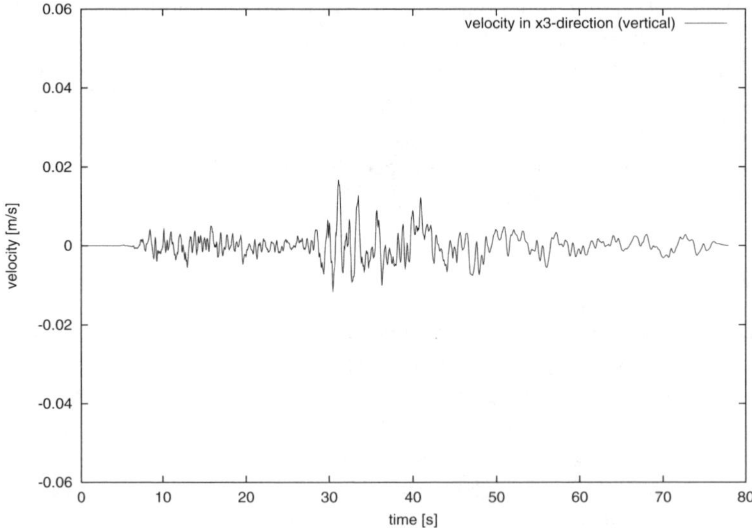

Fig. 7.2. Kobe earthquake data record (station Fukui), vertical velocity versus time

is very vulnerable against horizontal excitations. Nevertheless similar houses can be found in earthquake endangered areas. For detailed geometrical properties refer to Figs. 7.4 and 7.5. An isometric view of the building is shown in Fig. 7.6.

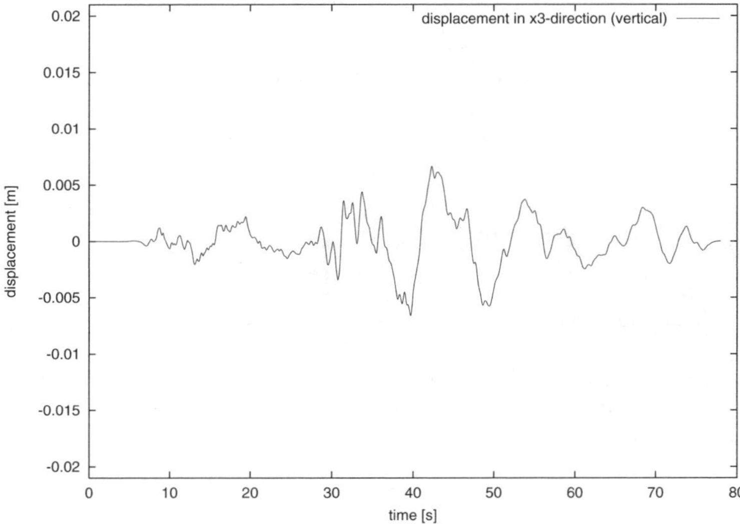

Fig. 7.3. Kobe earthquake data record (station Fukui), vertical displacement versus time

The simplest way of a seismic simulation of structures is a base excitation without taking the soil into account. For a more realistic simulation, soil-structure interaction has to be applied. For the first soil-structure interaction simulation, a homogeneous soil of middle dense sand is assumed with the characteristics: stiffness modulus $E_s = 100MPa$, resulting Young's modulus $E = 74.3MPa$ (calculated with relation (6.66)), Poisson's ratio $\nu = 0.3$, and density $\rho = 1900\frac{kg}{m^3}$.

The third model (and second soil-structure interaction example) consists of a layered soil , see Fig. 7.7. In this model, over middle dense sand (material parameters see above) a $3m$ thick clay layer (stiffness modulus $E_s = 10MPa$, resulting Young's modulus $E = 7.4MPa$, Poisson's ratio $\nu = 0.3$, and density $\rho = 2000\frac{kg}{m^3}$)[1] is placed.

The three simulation variants are denoted as "base excitation", "homogeneous soil" and "layered soil".

7.3.3 Discretisation

Fig. 7.8 depicts a coarse discretisation of the near-field (building and surrounding soil). Here, the building is discretised with 140 four-node linear plate elements which have a thickness of $0.3m$ and concrete material property (material parameters see above). For the soil-structure interaction models, these elements are directly coupled with eight-node isoparametric three-dimensional soil elements, where elastic material behaviour is assumed. For the homogeneous and layered soil $4 \times 4 \times 4 = 64$ elements represent the near-field. At the near-field/far-field interface these elements

---

[1] Soil parameters were chosen according to design rules for preliminary calculations after [120].

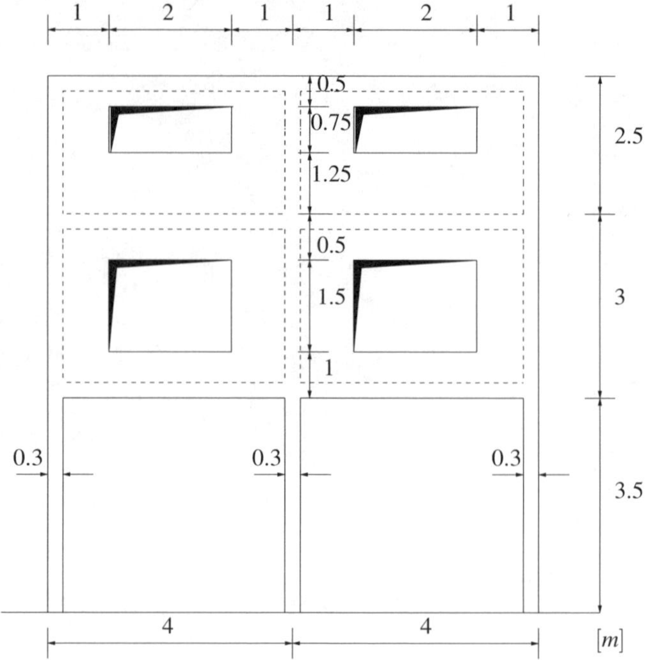

Fig. 7.4. Schematic front-view of discretised building

are coupled with 80 four-node scaled boundary finite elements. In order to verify the results, a simulation with a *refined* mesh is performed, as well. This refined mesh of the near-field is shown in Fig. 7.9. Now, the model contains 560 four-node linear plate elements, 512 eight-node three-dimensional soil elements and 320 four-node scaled boundary finite elements.

7.3.4 Results

For the simulation variant "homogenous soil", calculated displacements of points A, B and C (location of observation points see Fig. 7.8) for horizontal (x_1, x_2) and vertical (x_3) direction are plotted in Figs. 7.10 to 7.12. The simulation time of this calculation is reduced to $3s$, equivalent to 2000 time-steps $\Delta t = 1.5 \cdot 10^{-3}s$).

These results are gained with a "rigorous" and an "approximative" FE/SBFE procedure on the coarse and refined mesh. For the rigorous simulation, fully populated influence matrices and a direct calculation of the involved convolution integral are used, see Sec. 3.6. The approximative methodology (Sec. 3.7) leads to a reduction of non-locality in space and time. The reduction of globality in space is reached by introducing a zero-element threshold $\varepsilon_z = 10^{-4}$, resulting in sparse influence matrices. Furthermore, the utilisation of the proposed recursive algorithm eliminates the non-locality in time. A linear behaviour of the influence matrices can be observed

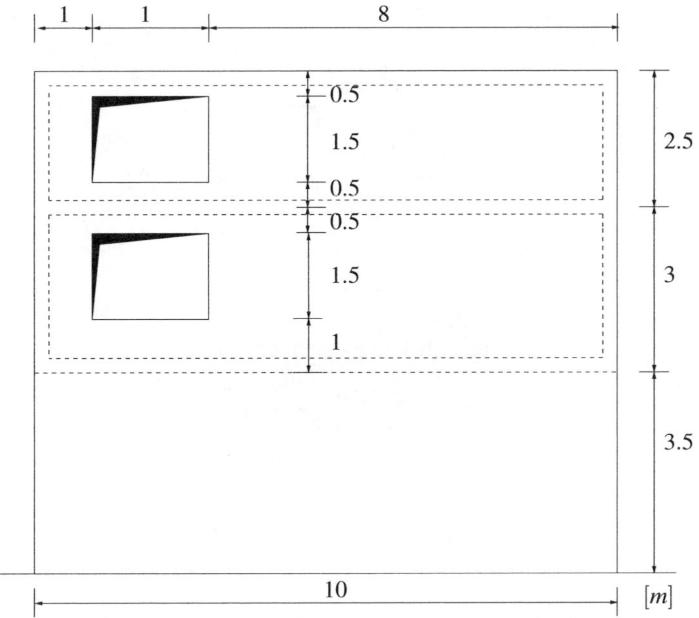

Fig. 7.5. Schematic side-view of discretised building

Fig. 7.6. Isometric views of discretised building

after 244 time-steps ($t_m = 0.366s$). Number of matrix entries per unit-impulse response matrix is reduced from $267^2 = 71289$ (total number of degrees of freedom of the near-field/far-field interface is 267) to 10266 for the coarse discretisation and from $1.02 \cdot 10^6$ to $1.12 \cdot 10^5$ for the refined mesh. Thus, the density (ratio of non-zero elements to total amount of matrix entries) of the influence matrices is reduced from 100% to 14.4% (coarse mesh) and to 11,0% (refined mesh), respectively.

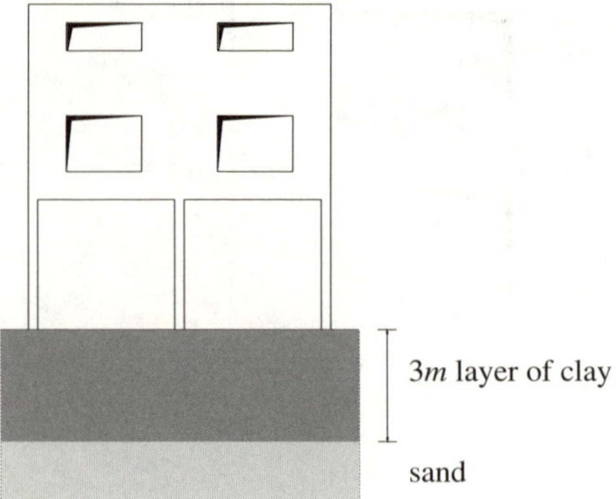

Fig. 7.7. Schematic cross section of layered soil

Calculated displacements of the direct solution (coarse mesh) show no deviation from approximative solutions (coarse and refined mesh), see Figs. 7.10 to 7.12.

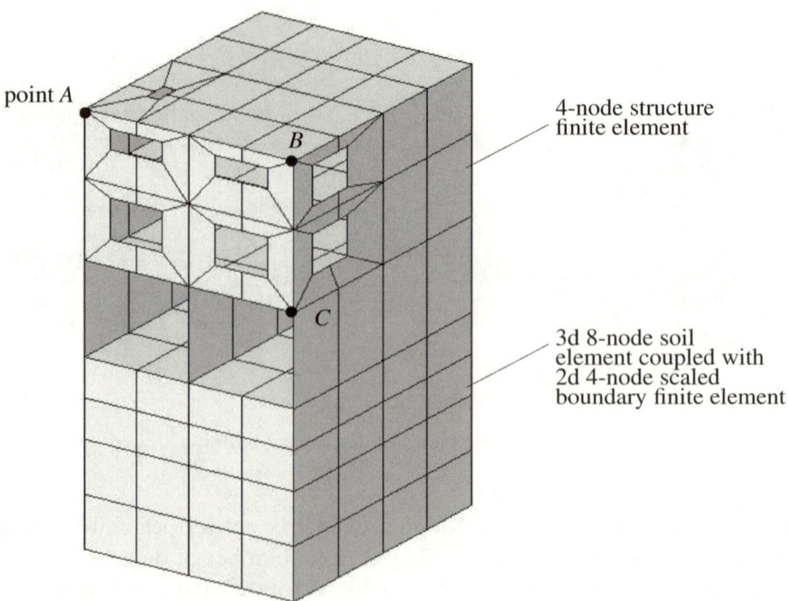

Fig. 7.8. Coarse discretisation of near-field and location of points *A*, *B* and *C*

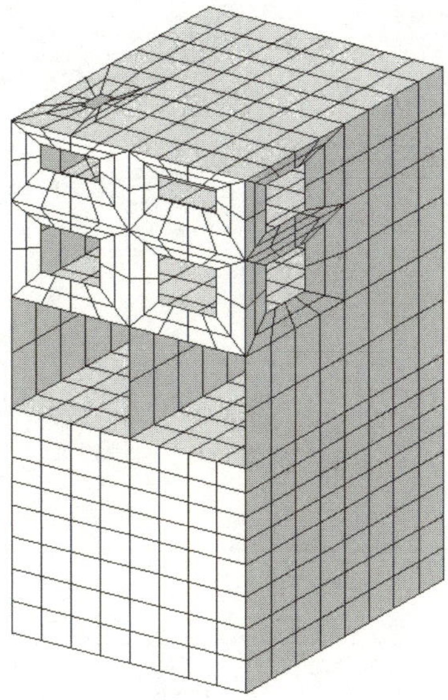

Fig. 7.9. Refined discretisation of near-field

For the first 2000 time-steps the approximative simulation needs only 1.7% of the CPU time of the direct solution. Fig. 7.14 shows calculated displacements of the building with the three soil models "base excitation", "homogeneous soil", and "layered soil" for different time-stations ($t_1 = 21s$, $t_2 = 36s$, and $t_3 = 60s$, see Fig. 7.13). The grey scale represents the magnitude of displacement: light marking corresponds to large displacements and dark to small ones. It is evident that for different soil models, the calculated displacements have a large deviation.

Figs. 7.15 to 7.17 display the calculated vertical displacements for observation points A, B and C for all soil models in dependency of the simulation time ($0 \leq t \leq 78s$). Here, especially the deviation of the base excitation model to the soil-structure interaction simulations is apparent. Due to the fact that for the base excitation simulation wave reflections occur at the excited finite element nodes, the results are distorted from a certain time-step on ($t \approx 30s$), see Figs. 7.15 to 7.17 and B.1 to B.6 of Appendix B.

For a simulation time of $78s$ it is required to calculate 52000 time-steps ($\Delta t = 1.5 \cdot 10^{-3}s$). With a linear behaviour of matrix entries from time-station t_m on, with $t_m = 0.366s$ ($m = 244$), the necessary operations reduce from $5.2 \cdot 10^{18}$ (equation (3.136)) to $4.9 \cdot 10^{16}$ (equation (3.137)), which results in a CPU time reduction of 99.1%.

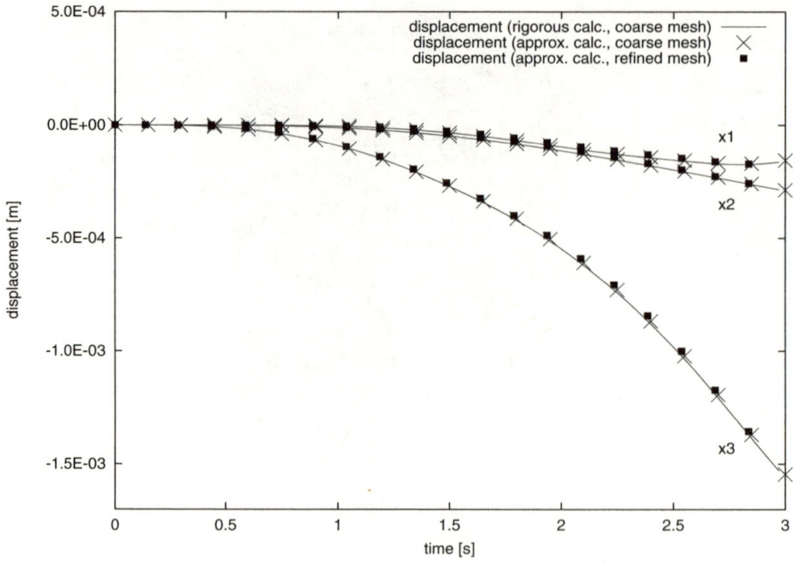

Fig. 7.10. Calculated horizontal (x_1, x_2) and vertical (x_3) displacements of point A versus time: comparison of direct and approximated solution, homogenous soil, coarse and refined mesh, for the first 2000 time-steps ($t \leq 3s$)

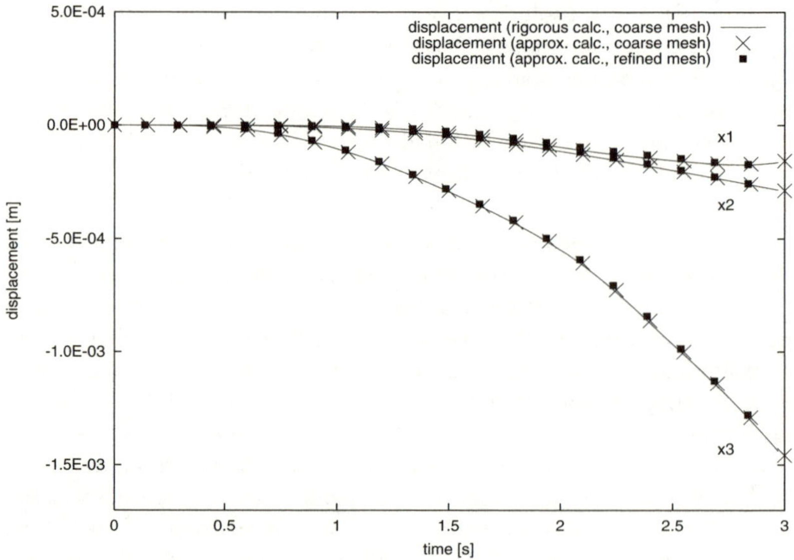

Fig. 7.11. Calculated horizontal (x_1, x_2) and vertical (x_3) displacements of point B versus time: comparison of direct and approximated solution, homogenous soil, coarse and refined mesh, for the first 2000 time-steps ($t \leq 3s$)

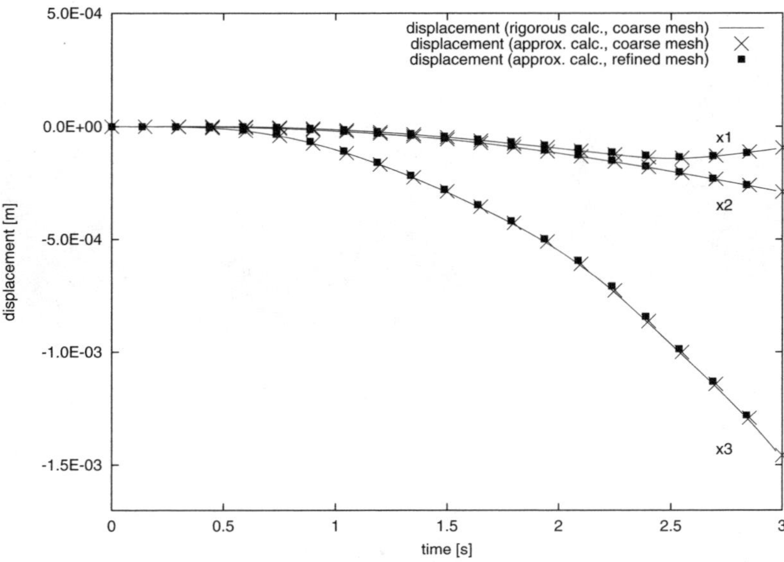

Fig. 7.12. Calculated horizontal (x_1, x_2) and vertical (x_3) displacements of point C versus time: comparison of direct and approximated solution, homogenous soil, coarse and refined mesh, for the first 2000 time-steps $(t \leq 3s)$

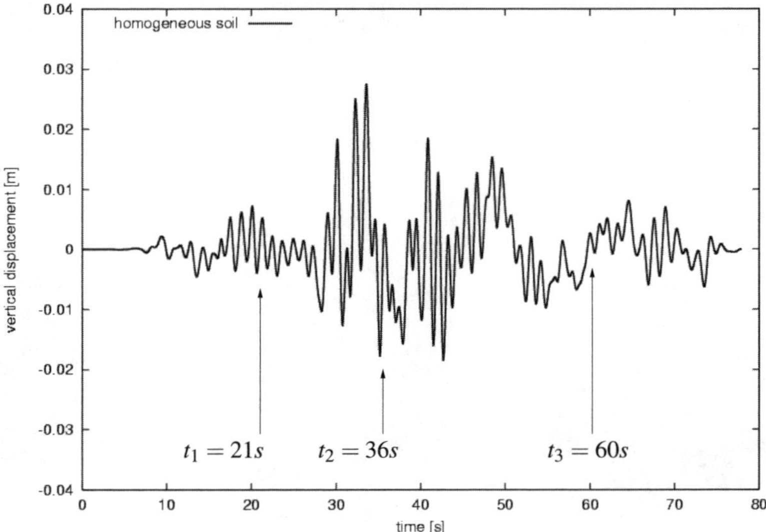

Fig. 7.13. Calculated vertical displacement versus time with definition of times t_i

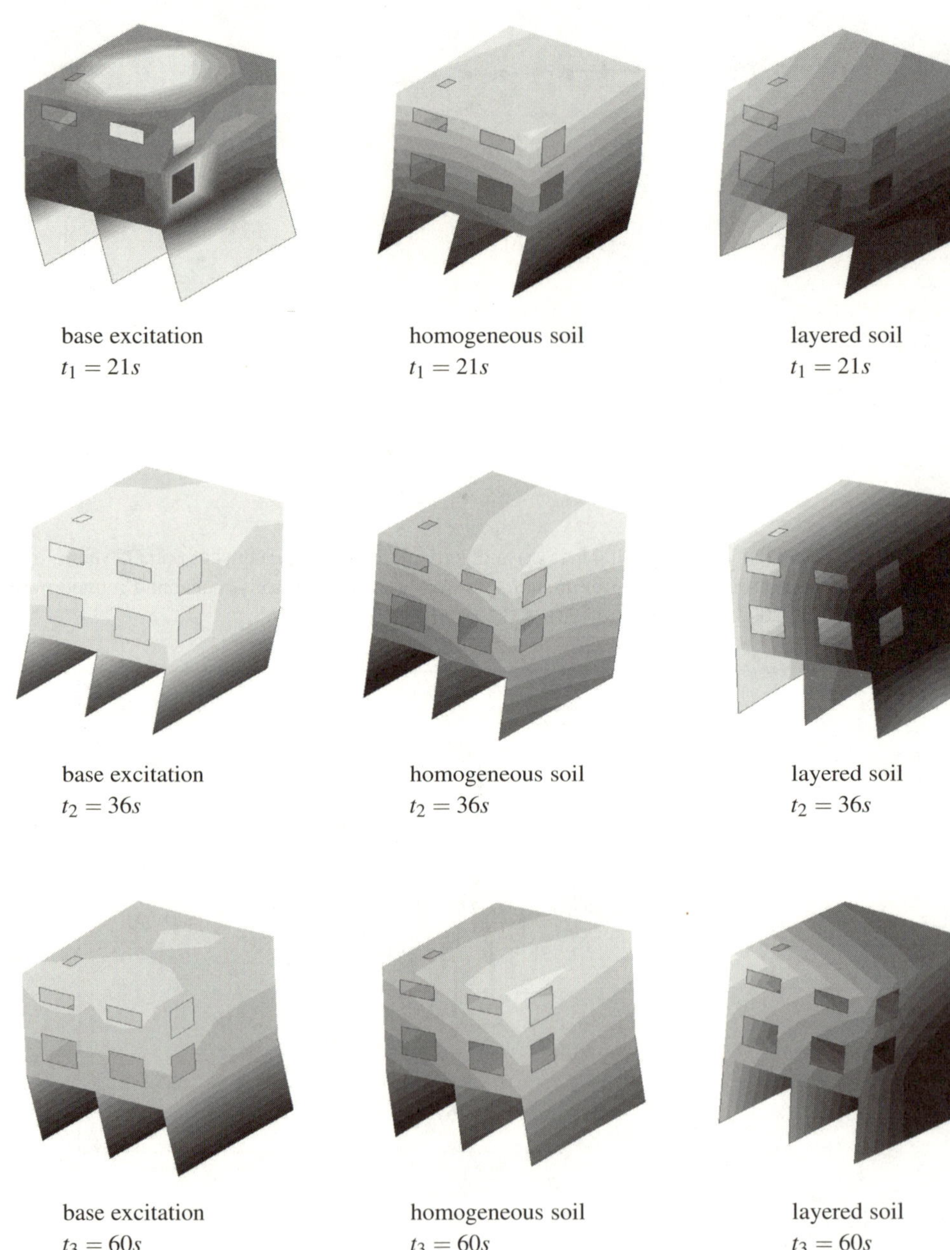

Fig. 7.14. Calculated displacement magnitudes of discretised building for time-station $t_1 = 21s$, $t_2 = 36s$, and $t_3 = 60s$: comparison of base excitation, homogeneous and layered soil

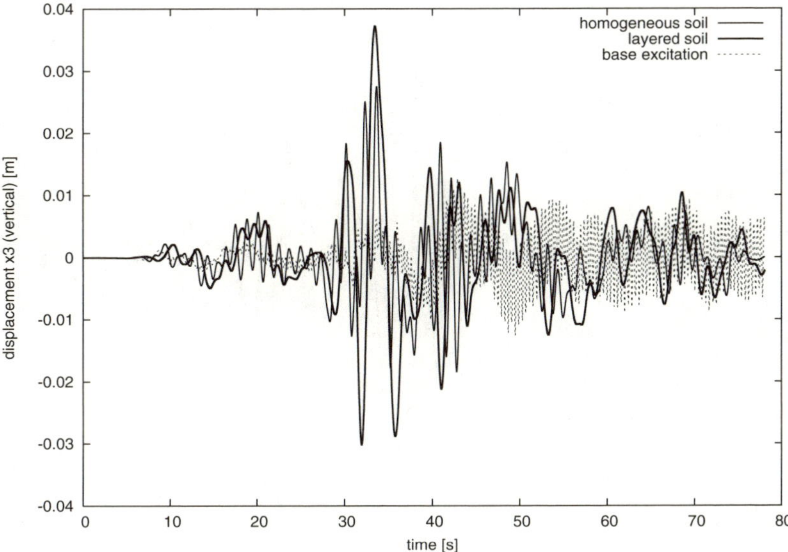

Fig. 7.15. Calculated vertical displacement of point *A* versus time: comparison of base excitation, homogeneous and layered soil

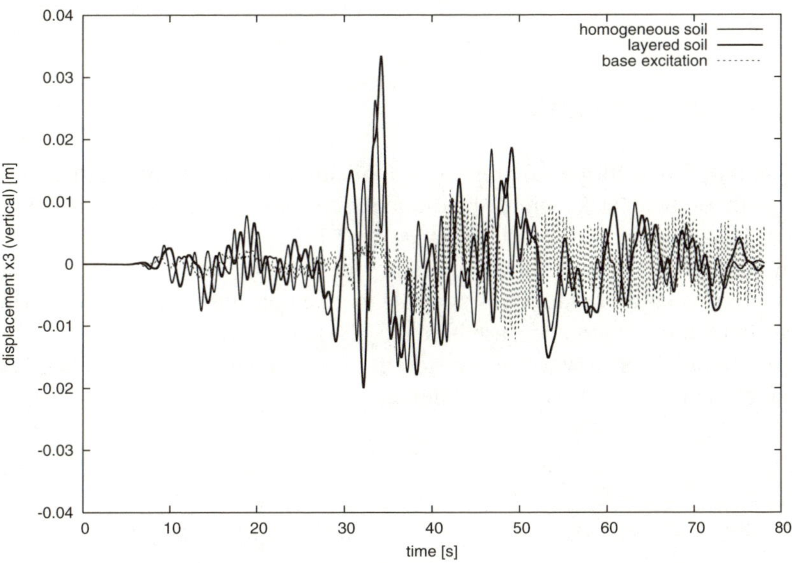

Fig. 7.16. Calculated vertical displacement of point *B* versus time: comparison of base excitation, homogeneous and layered soil

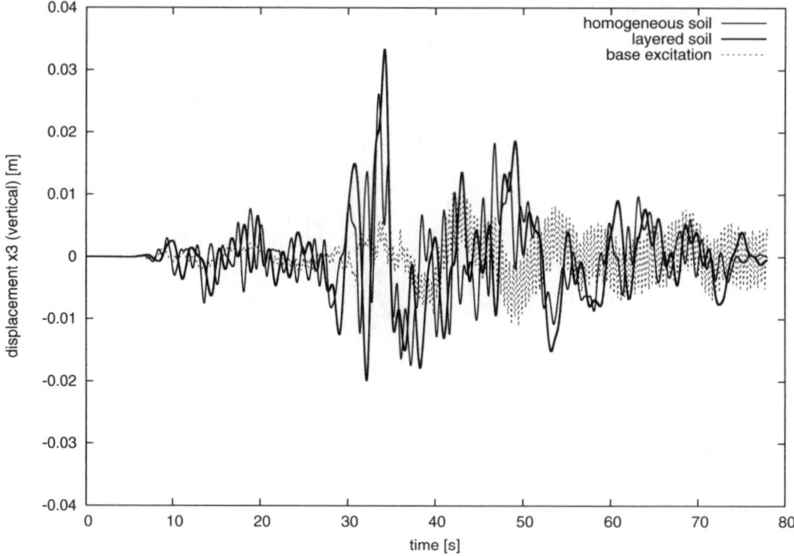

Fig. 7.17. Calculated vertical displacement of point C versus time: comparison of base excitation, homogeneous and layered soil

7.4 Concluding remarks

Focal point of this section is to present an application for the proposed methodology, while the theoretical background of earthquake engineering can be found, e.g., in the books of Coburn et al. [38] or Zienkiewicz et al. [152].

For the seismic simulation the efficiency of the introduced numerical model is proven. For this kind of model it is possible to take non-linear material behaviour of the near-field into account, because here the FEM is utilised.

Furthermore, it is shown that considering soil-structure interaction besides correct modelling of the soil layers is evident.

Conclusions and outlook

In this work, the theoretical basics and some important applications of a new numerical methodology for the analysis of wave propagation in infinite domains are presented. In principle, it consists of a near-field part mapped via the FEM and a far-field reproduced by a modified SBFEM. For comparison studies the BEM, or a coupled FE/BE technique is used. Applications of various engineering fields complete the introduced approach.

Theoretical background

By choosing the FEM for the near-field discretisation and the SBFEM as absorbing boundary conditions several advantages of the solution procedure are achieved: the FEM allows flexible handling of boundary conditions and geometries, and utilising the SBFEM reduces the spatial dimension by one, while no fundamental solution is needed. But the SBFEM – as it was originally developed by Wolf et al. [147] – introduces also some computational expensive characteristics to the numerical model. As it is formulated as *exact absorbing boundary condition*, it is non-local in time and space. The non-locality in time is the bottleneck for long simulation time analyses, and the globality in space for large near-field/far-field interfaces. The art of formulating a suitable boundary condition is to reduce the global nature of the absorbing boundary condition without accepting significant deviations to the exact solution.

Hence, several modifications are introduced. The time non-locality results in a convolution integral, which has to be solved. Therefore, the number of required operations grows quadratic in dependency on the simulation time for the direct "rigorous" formulation. In order to reduce these costs to a linear time dependency, a recursive algorithm is formulated.

Secondly, due to the space globality an immense storage consumption of the SBFEM results from fully populated unit-impulse acceleration influence matrices. While for each time-station a different influence matrix has to be recorded, the memory consumption is unacceptable for large near-field/far-field interfaces (which results in big influence matrices) or for long simulation times (equivalent to the recording of numerous matrices). Therefore, a storage reduction method is presented as well, which results in a banded, sparse structure, like stiffness matrices in finite elements. Thus, effective calculation techniques as special matrix vector multiplication algorithms become available.

The combined use of the recursive algorithm and the storage reduction method reduces the operating expense of the SBFEM significantly. The results of the presented benchmark examples prove the efficiency and correctness of the proposed method by comparing the results to a BE simulation.

Applications

Applications of the modified coupled FE/SBFE procedure include the simulation of wave propagation in fluids (scalar wave equation), a computation of an offshore wind turbine with consideration of the soil-structure interaction, and finally, a seismic analysis of a three-storey building.

Wave propagation in fluids is achieved in frequency and time domain. The frequency domain analyses inclose a coupled fluid/structure/fluid example, where the transmission of sound through a glass plate is simulated. With the proposed technique it is possible to predict the damping characteristics of separating components, such as windows with noise protection function. For an acoustical example in the time domain, the usage of a three-dimensional coupled FE/SBFE technique as application of hierarchical or $\mathcal{H}$-matrices [65, 66] is introduced. $\mathcal{H}$-matrices are evolved from panel clustering [67] which provided an alternative to the fast multipole method [62]. This method reduces the complexity of standard matrix operations from a quadratic to a quasi-linear cost in storage and computation time. The concept of $\mathcal{H}$-matrices is applied to the matrix vector multiplication (influence matrices $\times$ velocity vector) which arises during the evaluation of the convolution integral for time domain analysis. This approximation reduces the computational effort from quadratic dependency on degrees of freedom to linear dependency, without any loss of accuracy.

Secondly, the analysis of an offshore wind turbine in the time domain shows that the numerical model is suitable for dimensioning and fatigue analysis. Here, the multifaceted offshore environment with realistic wind- and wave loadings is considered. In addition, the structural dynamic behaviour including rotor dynamics enriches the calculation. To find a realistic representation of the subsoil as damper/dashpot system is difficult if not impossible, therefore the dynamic soil-structure interaction is modelled by applying a near-field representation of the surrounding soil. Again, a coupled FE/SBFE technique prevents wave reflection at the artificial boundary. A postprocessed analysis may include the computation of stress distribution in the tower system, where three-dimensional effects (like asymmetric openings in the tower) could be incorporated.

Thirdly, an earthquake engineering example is presented as application for the proposed methodology. For this earthquake engineering simulation, the efficiency of the introduced numerical model is proven. Again, the soil-structure interaction is taken into account, where the importance of the soil model is shown.

Outlook

An enhancement of the efficiency of the SBFEM could be reached by applying the above mentioned concept of $\mathcal{H}$-matrices on solving the Lyapunov and Riccati equations. These equations appear in the solution procedure for solving the non-linear partial differential equations of the SBFEM.

Due to the use of finite elements in the near-field, it is possible to take non-linear effects into account, suitable especially for soil-structure interaction simulations. Further simulations with non-linear material or structural behaviour should be executed. The relevance of the material formulation has to be investigated, because the use of non-linear material formulations causes some extra costs.

Introducing a porous media formulation as Biot's theory [22, 23, 24] is of considerable importance. A more complex constitutive model for saturated soil (e.g., as it appears as offshore wind turbine building ground) is able to guide the calculation to a more realistic one. Advantageous of the FE/SBFE model is the flexible procedure with which the underlying differential equations are solved. The model is ready for extension to a porous media formulation in frequency as well as in the time domain. All described improvements can be adopted without additional work.

A. Recorded horizontal accelerations, velocities and displacements during Kobe earthquake, Fukui station

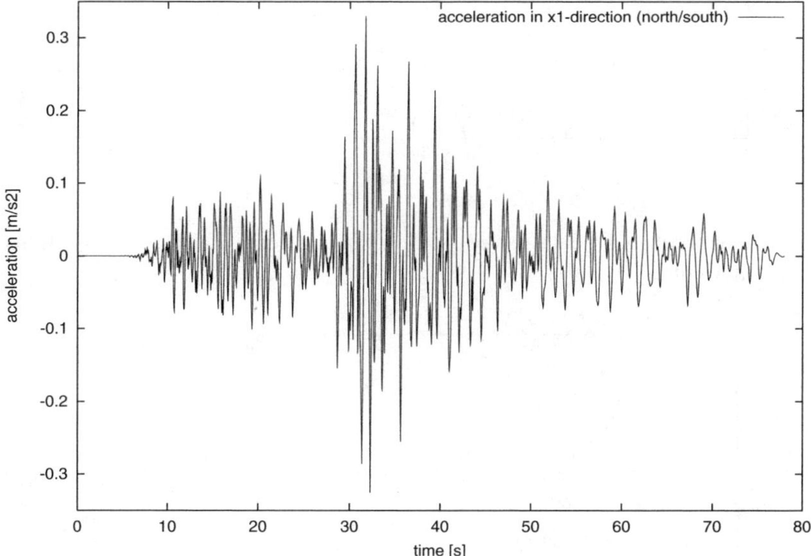

Fig. A.1. Kobe earthquake data record (station Fukui), horizontal acceleration (direction north/south) versus time

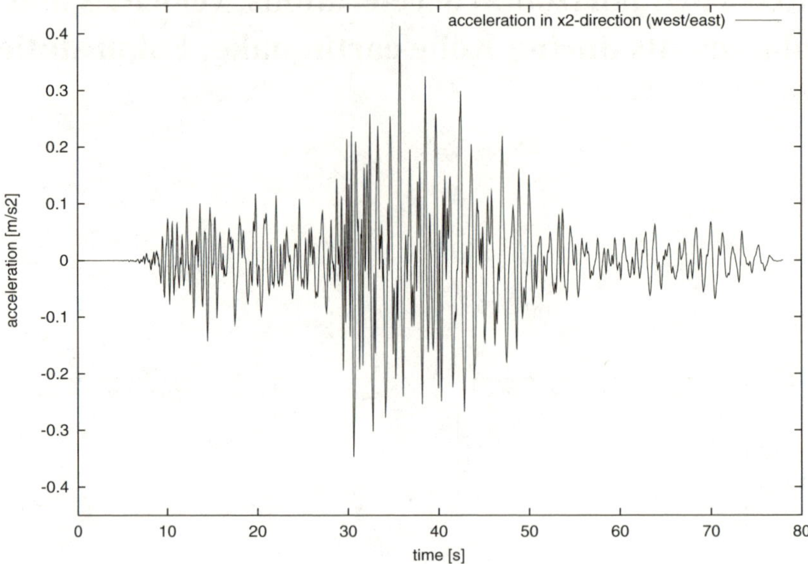

Fig. A.2. Kobe earthquake data record (station Fukui), horizontal acceleration (direction west/east) versus time

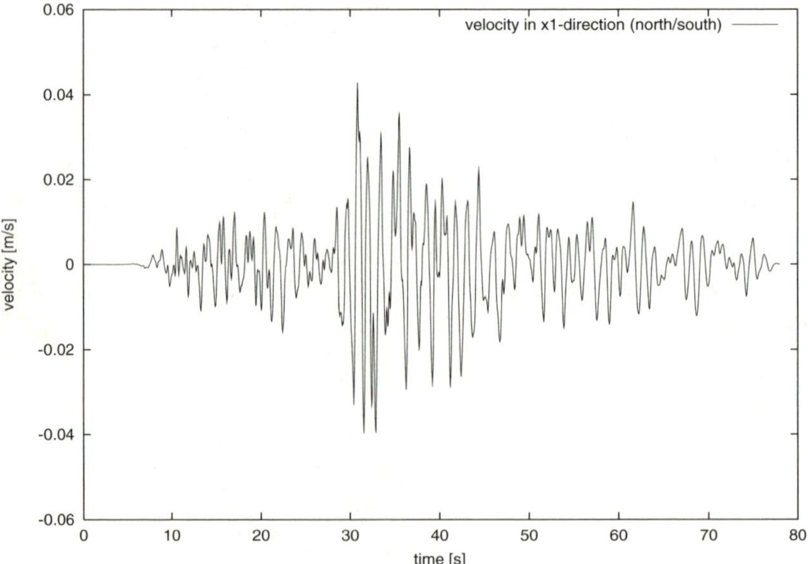

Fig. A.3. Kobe earthquake data record (station Fukui), horizontal velocity (direction north/south) versus time

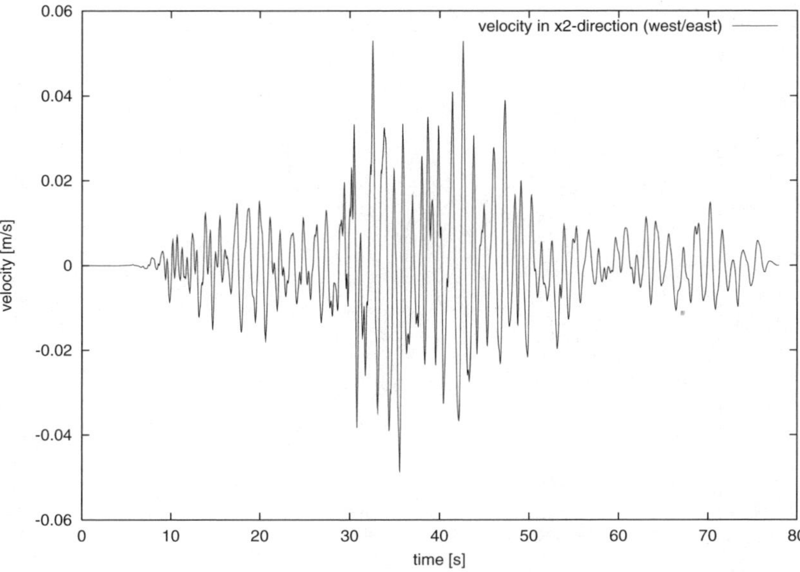

Fig. A.4. Kobe earthquake data record (station Fukui), horizontal velocity (direction west/east) versus time

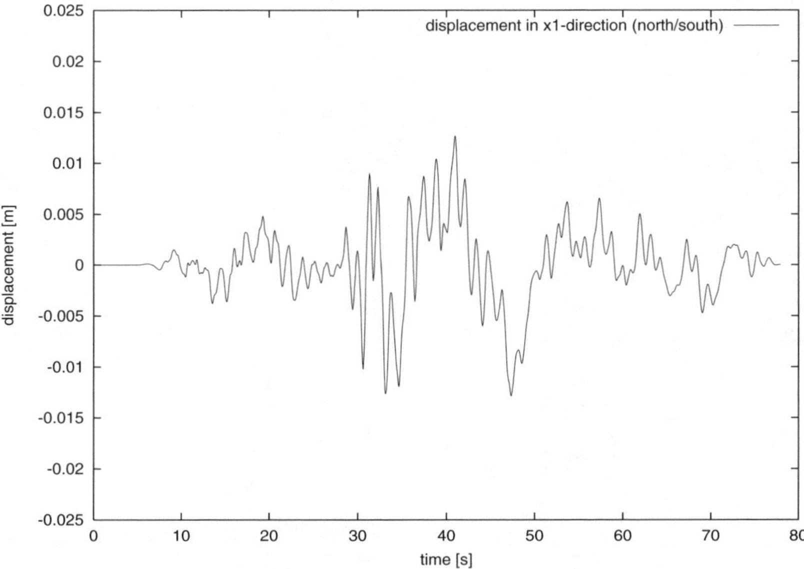

Fig. A.5. Kobe earthquake data record (station Fukui), horizontal displacement (direction north/south) versus time

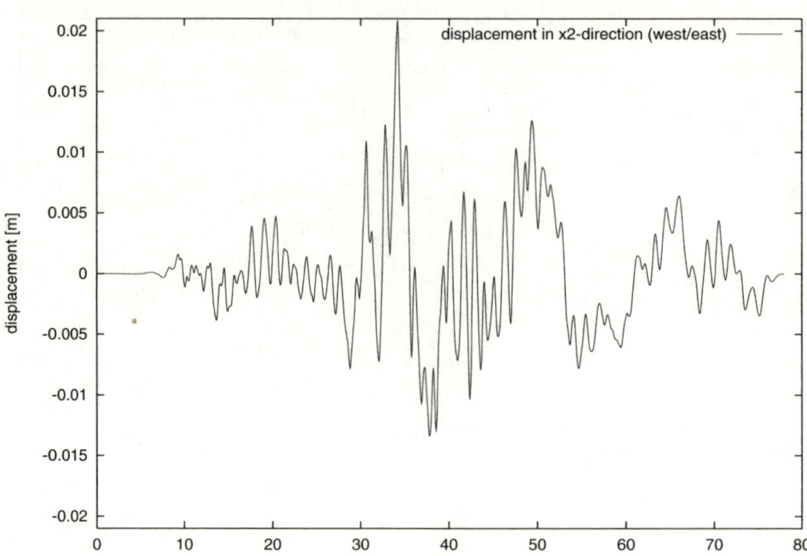

Fig. A.6. Kobe earthquake data record (station Fukui), horizontal displacement (direction west/east) versus time

B. Calculated displacements of earthquake excited building

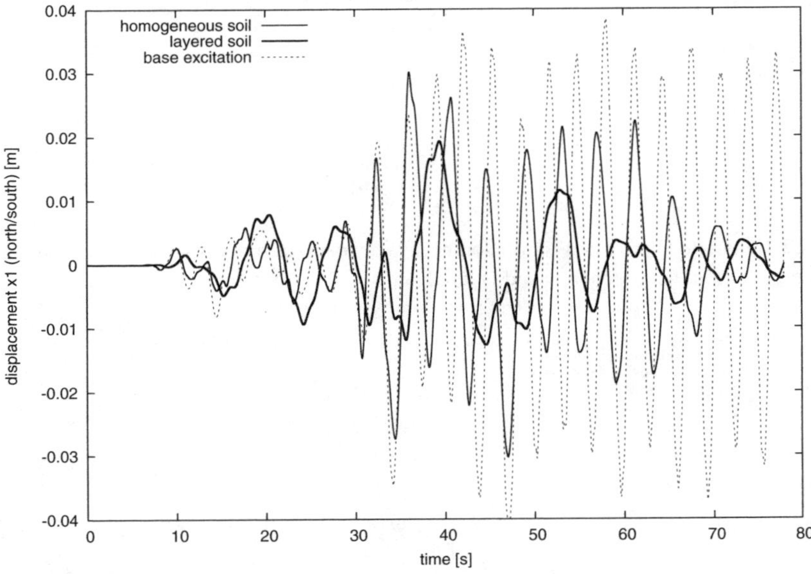

Fig. B.1. Calculated horizontal (north/south) displacement of point *A* versus time: comparison of base excitation, homogeneous and layered soil

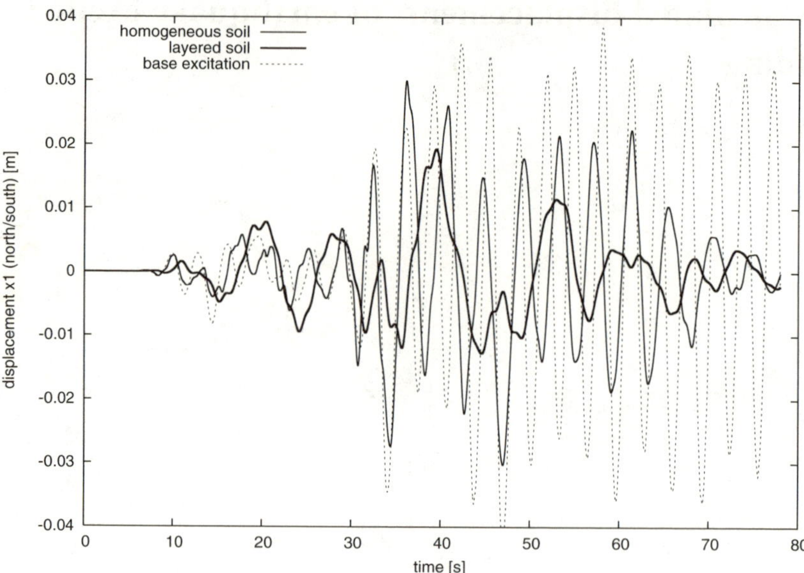

Fig. B.2. Calculated horizontal (north/south) displacement of point B versus time: comparison of base excitation, homogeneous and layered soil

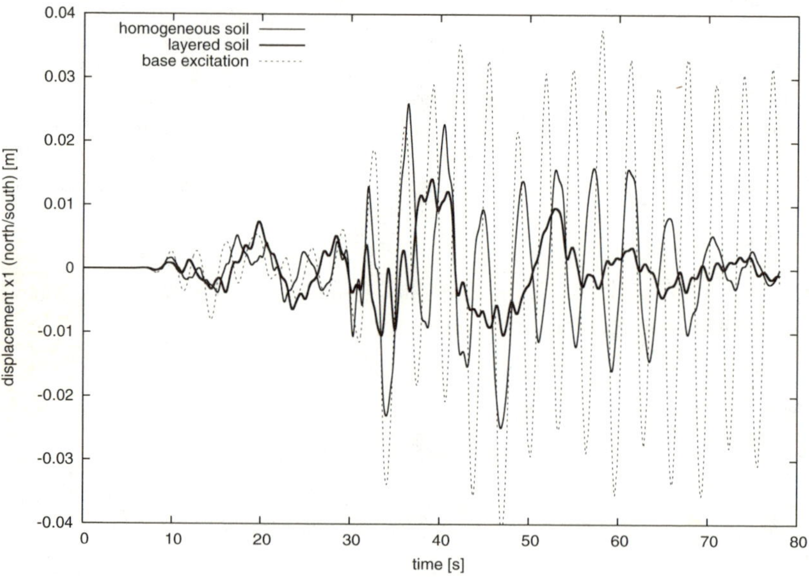

Fig. B.3. Calculated horizontal (north/south) displacement of point C versus time: comparison of base excitation, homogeneous and layered soil

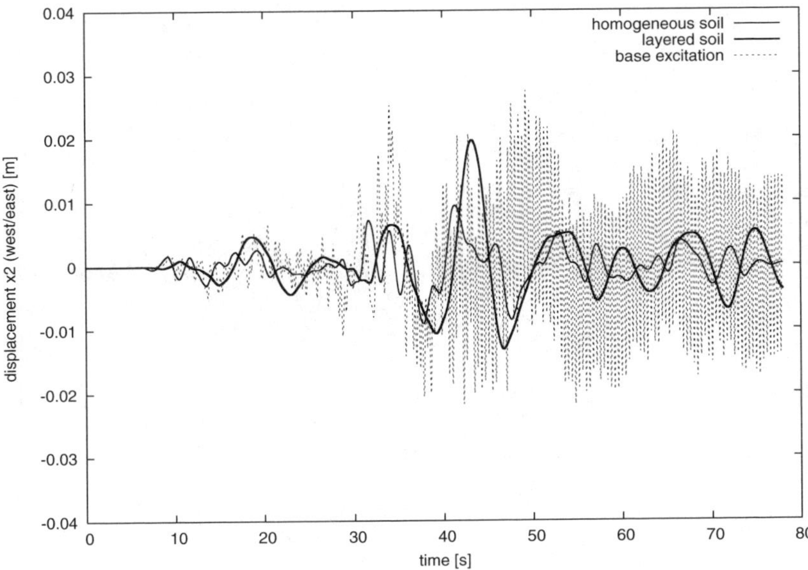

Fig. B.4. Calculated horizontal (west/east) displacement of point A versus time: comparison of base excitation, homogeneous and layered soil

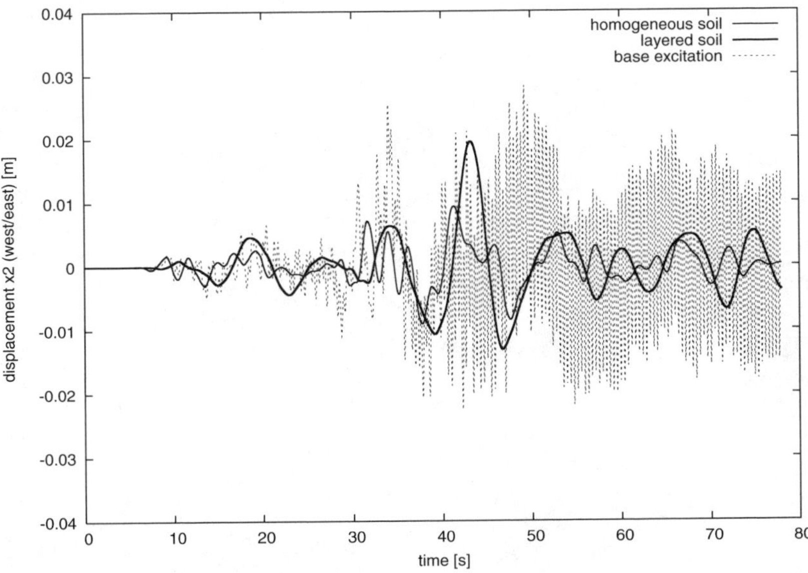

Fig. B.5. Calculated horizontal (west/east) displacement of point B versus time: comparison of base excitation, homogeneous and layered soil

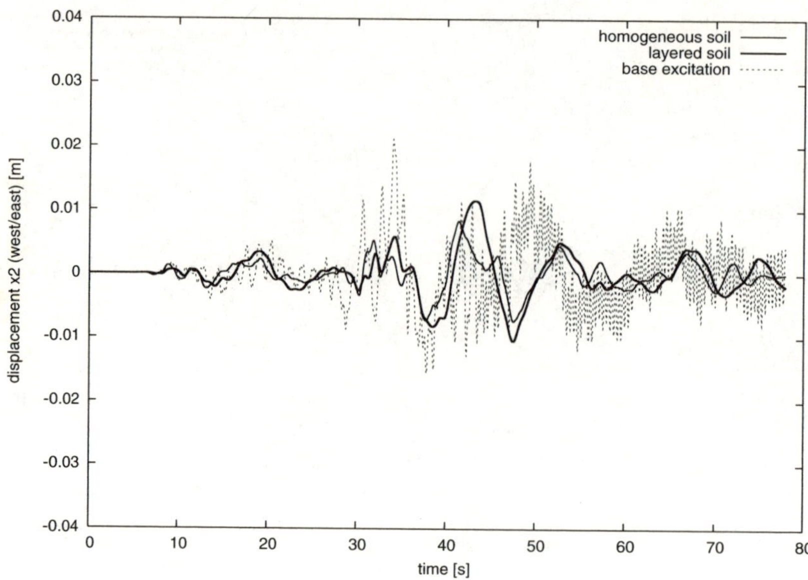

Fig. B.6. Calculated horizontal (west/east) displacement of point C versus time: comparison of base excitation, homogeneous and layered soil

References

1. Abarbanel, S.; Gottlieb, D.: A mathematical analysis of the PML method. *Journal of Computational Physics*, **134**(2), 357–363, 1997.
2. Airy, G.B.: On Tides and Waves. *Encyclopoedia Metropolitana*, **5**, 241–396, 1845.
3. Altmann, M.; Rickert, F.: Hydrogen production at offshore wind farms. *Offshore Wind Energy Special Topic Conference*, Brussels, Belgium, 2001.
4. amec: Blyth Offshore. http://www.amec.com/, London, UK, 2005.
5. American Petroleum Institute: Recommended Practice for Planning, Designing, and Constructing Offshore Platforms, 20th ed., RP-2A. Technical report, API, Dallas, USA, 1997.
6. Antes, H.: A boundary element procedure for transient wave propagations in two-dimensional isotropic elastic media. *Finite Elements in Analysis and Design*, **1**, 313–322, 1985.
7. Antes, H.: *Anwendungen der Methode der Randelemente in der Elastodynamik und der Fluiddynamik*. Teubner, Stuttgart, GER, 1988.
8. Antes, H.: Applications in environmental noise. In *Boundary Element Methods in Acoustics*. (Ciskowski, R.D.; Brebbia, C.A., Eds.), Springer, Berlin, GER, 225–260, 1991.
9. Antes, H.: A boundary element for the dead weight integral in 2-D elastodynamics BEM. *Mechanics Research Communications*, **19**, 273–278, 1992.
10. Antes, H.; Baaran, J.: Noise Radiation from Moving Surfaces. *Engineering Analysis with Boundary Elements*, **25**, 725–740, 2001.
11. Axelsson, O.: *Iterative solution methods*. Cambridge Univ. Press, Cambridge, UK, 1996.
12. Bartels, R. H.; Stewart, G. W.: Algorithm 432: Solution of the matrix equation AX + XB = C. *Communications of the ACM*, **15**, 820–826, 1972.
13. Bathe, K.-J.: *Finite Element Procedures in Engineering Analysis*. Prentice-Hall, Englewood Cliffs, USA, 1982.
14. Bathe, K.-J.: *Finite Element Procedures*. Prentice-Hall, Englewood Cliffs, USA, 1996.
15. Beer, G.: *Programming the Boundary Element Method*. Wiley and Sons, Chichester, UK, 2001.
16. Beer, G.; Golser, H.; Jedlitschka, G.; Zacher, P.: Coupled finite element/boundary element analysis in rock mechanics – industrial applications. In *Rock Mechanics for Industry*. (Amadei, B.; Kranz, R.L.; Scott, G.A.; Smeallie, P., Eds.), Balkema, Rotterdam, NL, 1999.
17. Beer, G.; Watson, J.O.: Infinite Boundary Elements. *International Journal for Numerical Methods in Engineering*, **28**, 1233–1247, 1989.
18. Beer, G.; Watson, J.O.: *Introduction to Finite and Boundary Element Methods for Engineers*. Wiley and Sons, Chichester, UK, 1992.
19. Berenger, J.P.: A perfectly matched layer for the absorption of electromagnetic waves. *J. Comput. Phys.*, **114**(2), 195–200, 1994.

20. Bettess, P.: *Infinite Elements*. Penshaw Press, 1992.

21. Bielak, J.; Christiano, P.: On the effective seismic input for non-linear soil-structure interaction systems. *Earthquake Engineering and Structural Dynamics*, **12**, 107–119, 1984.

22. Biot, M.A.: General Theory of Three-Dimensional Consolidation. *Journal of Applied Physics*, **12**, 155–164, 1941.

23. Biot, M.A.: Theory of Elasticity and Consolidation for a Porous Anisotropic Solid. *Journal of Applied Physics*, **26**, 182–185, 1955.

24. Biot, M.A.: Theory of Propagation of Elastic Waves in a Fluid-Saturated Porous Solid. I. Low-Frequency Range, II. Higher Frequency Range. *Journal of the Acoustical Society of America*, **28**(2), 168–191, 1956.

25. Bolt, B. A.: *Earthquakes*. WH Freeman, Berkeley, USA, 4th edition, 1999.

26. Bonnet, M.: *Boundary integral equation methods for solids and fluids*. Wiley and Sons, Chichester, UK, 1995.

27. Bonnet, M.; Maier, G.; Polizzotto, C.: Symmetric Galerkin Boundary Element Methods. *Applied Mechanics Reviews*, **51**, 669–704, 1998.

28. Börm, S.; Grasedyck, L.: . HLIB — a library for $\mathcal{H}$- and $\mathcal{H}^2$-matrices, 2004. http://www.hlib.org.

29. Börm, S.; Grasedyck, L.; Hackbusch, W.: Introduction to hierarchical matrices with applications. *Eng. Anal. Boundary Elements*, **27**, 405–422, 2003.

30. Börm, S.; Grasedyck, L.; Hackbusch, W.: Hierarchical Matrices. Lecture Notes for Winterschool on $\mathcal{H}$-Matrices, 2004. http://www.mis.mpg.de/scicomp/Fulltext/WS_HMatrices.pdf.

31. Bundesanstalt für Seeschifffahrt und Hydrographie: Offshore Foundations: Monopile. http://www.bsh.de/de/Meeresdaten/Beobachtungen/Eis/1952.jsp, Hamburg, GER, 2005.

32. Cederfeldt, L.: A Finite Element Approach to Two-Dimensional Sound Radiation and a Discussion of Sound Insulation of Corrugated Plates. *Report 59, Division of Building Technology, Lund Institute of Technology*, 1975.

33. Cen, Z.Z.; Swoboda, G.; Du, Q.H.: Damage analysis by boundary element method. *Acta Mechanica Solida Sinica*, **7**, 245–255, 1994.

34. Chakrabarti, S.K.: *Hydrodynamics of Offshore Structures*. Springer, London, UK, 1987.

35. Charnock, H.: Wind stress on a water surface. *Quarterly Journal of the Royal Meteorological Society*, **81**, 639–640, 1955.

36. Cheng, P.W.: *A Reliability Based Design Methodology for Extreme Responses of Offshore Wind Turbines*. DUWIND Delft University Wind Energy Research Institute, Delft, NL, 2002.

37. Clasen, D.; Langer, S.; Schanz, M.: Efficient Simulation of Sound Insulation in Building Acoustics. *Proc. of the Joint Congress CFA/DAGA04*, SFA DEGA, 303–304, 2004.

38. Coburn, A.; Spence, R.: *Earthquake Protection*. Wiley and Sons, Chichester, UK, 2nd edition, 2002.

39. Coifman, R.; Rokhlin, V.; Wandzura, S.: The Fast Mulitpole Method for the Wave Equation – A Pedestrian Prescritption. *IEEE Antennas and Propagation Magazine*, **35**(3), 7–12, 1993.

40. Cremer, L.; Heckl, M.: *Körperschall*. Springer-Verlag, Berlin, 1996.

41. Cuthill, E.; McKee, J.: Reducing the bandwidth of sparse symmetric matrices. *24th Nat. Conf.* ACM, 157–172, 1969.

42. Danish Wind Industry Association: Offshore Foundations: Monopile. http://www.windpower.org/en/tour/rd/monopile.htm, 2005. Copenhagen, DK.

43. Demkowicz, L.; Gerdes, K.: Convergence of the infinite element methods for the Helmholtz equation in separable domains. *Numerische Mathematik*, **79**, 11–42, 1998.

44. Dominguez, J.: *Boundary Elements in Dynamics*. Computational Mechanics Publications, Southampton, UK, 1993.

45. Dragt, J.B.: Atmospheric Turbulence Characteristics in the Rotating Frame of Reference of a WECS. In *Rotor. Proc. EWEC 1990*. (Stephens, H.S.; Associates, Eds.), 274–278, 1990.

46. (ed.), O. Steinbach: Hauptseminar Hierarchische Matrizen. Berichte aus dem Institut für Angewandte Analysis und Numerische Simulation, 2004. http://preprints.ians.uni-stuttgart.de.

47. El Naggar, M.H.: Seismic Response of Structures with Underground Storeys. Technical report, Institute of Catastrophic Loss Reduction, Toronto, CD, 2003.

48. Engineering Sciences Data Unit International: Data Item 74031: Characteristics of atmospheric turbulence near the ground. Part II: Single point data for strong winds. Technical report, ESDU, Mc Lean, USA, 1974.

49. Eringen, A.C.; Suhubi, E.S.: *Elastodynamics, Volume II, Linear Theory*. Academic Press, New York, USA, 1975.

50. Estorff, O. v.: *Boundary Elements in Acoustics - Advances & Applications*. WIT Press, 2000.

51. Estorff, O. v.; Prabucki, M. J.: Dynamic response in the time domain by coupled boundary and finite elements. *Comp. Mech.*, **6**, 33–46, 1990.

52. Etube, L.S.: *Fatigue and fracture mechanics of offshore structures*. Professional Engineering Publications, London, UK, 2001.

53. Fischer, M.; Gaul, L.: Application of the fast multipole BEM for structural-acoustics simulations. Technical report, Univ. of Stuttgart, Stuttgart, GER, 2003. SFB 404.

54. Gasch, R.: *Windkraftanlagen*. Teubner, Stuttgart, GER, 2005.

55. Gaul, L.; Kögel, M.; Wagner, M.: *Boundary Element Methods of Engineers and Scientists*. Springer, Berlin, GER, 2003.

56. Geography Resources: The Kobe Earthquake. http://www.zephryus.demon.co.uk/education/geog/tectonics/kobe.html, London, UK, 2005.

57. Germanischer Lloyd: *Rules for Regulations IV – Non-Marine Technology, Part 2 Offshore Wind Energy, Regulations for the Certification of Offshore Wind Energy Conversion Systems*. Germanischer Lloyd, Hamburg, GER, 1995.

58. Giljohann, D.; Bittner, M.: The Three-Dimensional DtN Finite Element Method for Radiation Problems of the Helmholtz Equation. *Journal of sound and vibration*, **212**(3), 383–394, 1998.

59. Givoli, D.: Non-reflecting boundary conditions. *Journal of Computational Physics*, **94**(1), 1–29, 1991.

60. Givoli, D.: High-order local non-reflecting boundary conditions: a review. *Wave Motion*, **39**(4), 319–326, 2004.

61. Gould, P.L.: *Introduction to Linear Elasticity*. Springer, Berlin, GER, 1994.

62. Greengard, L.: Fast algorithms for classical physics. *Science*, **265**, 909–914, 1994.

63. Haas, M.; Kuhn, G.: Mixed-dimensional, symmetric coupling of FEM and BEM. *Eng. Anal. Boundary Elements*, **27**, 575–582, 2003.

64. Hackbusch, W.: A Sparse Matrix Arithmetic based on H-Matrices. Part I: Introduction to H-Matrices. *Computing*, **62**, 89–108, 1998.

65. Hackbusch, W.: A sparse Matrix Arithmethic based on $\mathcal{H}$-Matrices. Part I: Introduction to $\mathcal{H}$-Matrices. *Computing*, **62**, 89–108, 1999.

66. Hackbusch, W.; Khoromskij, B.N.: A Sparse $\mathcal{H}$-Matrix Arithmetic. Part II: Application to Multi-Dimensional Problems. *Computing*, **64**(1), 21–47, 2000.

67. Hackbusch, W.; Nowak, Z.P.: On the Fast Matrix Multiplication in the Boundary Element Method by Panel Clustering. *Numerische Mathematik*, **54**, 463–91, 1989.

68. Hafenbautechnische Gesellschaft e.V. und Deutsche Gesellschaft für Geotechnik e.V.: *Empfehlungen des Arbeitskreises "Ufereinfassungen" Häfen und Wasserstraßen EAU 2004*. Ernst und Sohn, Berlin, 2005.

69. Hagstrom, T.: Radiation boundary condtions for the numerical simulation of waves. *Acta Numerica*, **8**, 47–106, 1999.

70. Hampel, S.; Langer, S.; Antes, H.: Representing outdoor sound propagation effects with a BEM model. *Proceedings of the Joint Congress CFA/DAGA04*, 2004.

71. Hansen, K.; Larsen, G.; Courtney, M.: Database on wind characteristics. *Proceedings of the European Wind Energy Conference*, European Wind Energy Association, EWEA, Copenhagen, DK, 2001.

72. Harari, I.; Slavutin, M.; Turkel, E.: Analytical and numerical studies of a finite element PML for the Helmholtz equation. *Journal of Computational Acoustics*, **8**(1), 121–137, 2000.

73. Harris, R.I.; Deaves, D.M.: The structure of strong winds. *The wind engineering in the eighties. Proceedings of CIRIA Conference 12/13*, London, UK, 4.1–4.93, 1980. Construction Industry, Research and Information Association.

74. Hasselmann, K. et al.: Measurements of Wind-Wave Growth and Swell Decay during the Joint North Sea Wave Project (JONSWAP). Technical report, Detusches Hydrographisches Institut, Hamburg, GER, 1973.

75. Hilber, H.M.; Hughes, T.J.R.; Taylor, R.L.: Improved numerical dissipation for time integration algorithms in strucutral dynamics. *Earthquake Engng. and Struct. Dyn.*, **5**, 283–292, 1977.

76. Hughes, T.J.R.: *Analysis of transient algorithms with particular reference to stability behaviour*, 67–155. North-Holland, 1983.

77. Hughes, T.J.R.: *The Finite Element Method: Linear Static and Dynamic Finite Element Analysis*. Prentice-Hall, Englewood Cliffs, USA, 1987.

78. Ihlenburg, F.: *Finite Element Analysis of Acoustic Scattering*. Springer Verlag, Berlin, GER, 1998.

79. Insitute for Energy and Environment, University of Strathclyde: Offshore Wind Turbines.
http://www.esru.strath.ac.uk/EandE/Web_sites/98-9/offshore/wind/wintr.htm, Glasgow, UK, 2005.

80. Institut für Luft- und Raumfahrt, ILR: Vindeby Offshore Wind Power Plant.
http://www.ilr.tu-berlin.de/WKA/windfarm/offshore.vindeby.html, Berlin, GER, 2005.

81. Jeans, R.A.; Mathews, I.C.: A unique coupled boundary element/finite element method for the elastoacoustic analysis of fluid-filled thin shells. *Journal of the Acoustical Society of America*, **94**(6), 3473–3479, 1993.

82. Karabalis, D.L.; Rizos, D.C.: Dynamic analysis of 3-D foundations. In *Boundary Element Techniques in Geomechanics*. (Manolis, G.D.; Davies, T.G., Eds.), Elsevier, London, UK, 1993.

83. Kopuz, S.; Ünlüsoy, Y.S.; Caliskan, M.: Integrated FEM/BEM approach to the dynamic and acoustic analysis of plate structures. *Engineering analysis with boundary elements*, **17**, 269–277, 1996.

84. Koslowski, G.: Der Eiswinter 1986/87 im deutschen Küstengebiet zwischen Ems und Trave. *Deutsche Hyrdographische Zeitung*, **40**, 117–123, 1987.

85. Kramer, S. L.: *Geotechnical Earthquake Engineering*. Prentice-Hall, Englewood Cliffs, USA, 1996.

86. Kühn, M.: *Dynamics and Design Optimisation of Offshore Wind Energy Conversion Systems*. DUWIND Delft University Wind Energy Research Institute, Delft, NL, 2001.

87. Langer, S.: *Schalltransmission durch Isolierverglasung*, Vol. 41, *Braunschweiger Schriften zur Mechanik*. Mechanik-Zentrum, Tech. Univ. of Braunschweig, Braunschweig, GER, 2001.

88. Langer, S.; Antes, H.: Analyses of Sound Transmission through Windows by coupled Finite and Boundary Element Methods. *Acta Acustica united with Acustica*, **89**, 78–85, 2003.

89. Lehmann, L.: Application of a Coupled Finite Element/Scaled Boundary Finite Element Procedure to Acoustics. *Coupled Problems 2005*, Santorini, Griechenland, 2005, European Community on Computational Methods in Applied Sciences (ECCOMAS).

90. Lehmann, L.: An effective finite element approach for soil-structure analysis in the time-domain. *Structural Engineering and Mechanics*, **21**, 437–450, 2005.

91. Lehmann, L.; Antes, H.; Schanz, M.: Transient Analysis of Soil-Structure Interaction Problems: An Effective FEM/SBFEM Approach. In *Advanced Numerical Analyses of Solids and Structures, and Beyond*. (G.A. Holzapfel, G. Rechard,W. Moser, Ed.), Verlag der Technischen Universität Graz, Graz, A, 99–116, 2004, Institute for Structural Analysis.

92. Lehmann, L.; Rüberg, T.: Application of Hierarchical Matrices to the simulation of wave propagation in fluids. *Communications in Numerical Methods in Engineering*, **22**, 489–503, 2006.

93. Lubich, C.: Convolution Quadrature and Discretized Operational Calculus. I. *Numerische Mathematik*, **52**, 129–145, 1988.

94. Lubich, C.: Convolution Quadrature and Discretized Operational Calculus. II. *Numerische Mathematik*, **52**, 413–425, 1988.

95. Mansur, W.J.: *A Time Stepping Technique to solve Wave Propagation Problems Using the Boundary Element Method*. dissertation, University of Southampton, 1983.

96. Mansur, W.J.; Brebbia, C.A.: Transient elastodynamics using a time-stepping technique. In *Boundary Elements*. (Brebbia, C.A.; Futagami, T.; Tanaka, M.; Tanaka, M., Eds.), Springer, Berlin, GER, 677–698, 1983.

97. Mantic, V.: A new formula for the C-matrix in the Somigliana Identity. *Journal of Elasticity*, **33**, 191–201, 1993.

98. Matthies, H.G.; Nath, C.; Schellin, T.E.; Garrad, A.D.; Wastling, M.A.; Quarton, D.C.; Wei, J.; Scherweit, M.; Siebert, T.: *Study of Offshore Wind Energy in the EC*. Verlag natürliche Energie, 1995.

99. Meyer, M.: *Reduktionsmethoden zur Simulation des aerolastischen Verhaltens von Windkraftanlagen*. dissertation, Tech. Univ. of Braunschweig, 2002.

100. Mindlin, R.D.: Influence of Rotatory Inertia and Shear on Flexural Motions of Isotropic Elastic Plates. *Journal of Applied Mechanics ASME*, **18**, 31–38, 1951.

101. Morison, J.R.; O'Brien, M. P.; Johnson, J. W.; Schaaf, S.A.: The force exerted by surface waves on piles. *AIME Petroleum Transactions*, 149–157, 1950.

102. Neumann, G.; Pierson, W.J.: *Principles of Physical Oceanography*. Prentice-Hall, Englewood Cliffs, USA, 1963.

103. Newland, D.E.: *An Introduction to Random Vibrations, Spectral and Wavelet Analysis*. Elsevier Science, Philadelphia, USA, 1997.

104. Nishimura, N.: Fast multipole accelerated boundary integral equation methods. *Applied Mechanics Reviews*, **55**, 299–324, 2002.

105. OWE – Offshore Windenergy Europe: The offshore wind energy website. http://www.offshorewindenergy.org, Delft Technical University,NL, 2005.

106. Pacific Earthquake Engineering Research Center: Strong motion database. http://peer.berkeley.edu/smcat/, Berkeley, USA, 2005.

107. Panofsky, H.A.: The atmospheric boundary layer below 150m. *Annual Review of Fluid Mechanics*, **6**, 147–177, 1974.

108. Paris, F.; Canas, J.: *Boundary Element Method: Fundamentals and Applications*. Oxford University Press, Oxford, UK, 1997.

109. Patel, M.H.: *Dynamics of Offshore Structures*. Butterworths, London, UK, 1989.

110. Peyton, H.R.: Ice and marine structures. Technical report, Ocean Industry Magazine, Houston, USA, 1968.

111. Pierson, W.J.; Moskowitz, L.: A proposed spectral form for fully developed wind seas based on the similarity theory of S.A. Kitaigorodskii. *J. Geophys. Res.*, **69**, 5181–5190, 1964.

112. Poulos, H.G.: *Grundbau Taschenbuch Teil 1*, Chapter Spannungen und Setzungen im Boden. Ernst und Sohn, Berlin, GER, 6th edition, 2001.

113. Poulos, H.G.; Davis, E.H.: *Elastic Solutions for Soil and Rock Mechanics*. Wiley and Sons, Chichester, UK, 1974.

114. Price, W.G.; Bishop, R.E.D.: *Probalistic Theory of Ship Dynamics*. Chapman and Hall, London, UK, 1974.

115. Rizos, D.C.; Wang, Z.: Coupled BEM-FEM solutions for direct time domain soil-structure interaction analysis. *Eng. Anal. Boundary Elements*, **26**, 877–888, 2002.

116. Schanz, M.: *Wave Propagation in Viscoelastic and Poroelastic Continua*. Springer Verlag, Berlin, GER, 2001.

117. Schlichting, H.: *Boundary-Layer Theory*. McGraw-Hill, New York, USA, 1968.

118. Shinozuka, M.; Jan, C.M.: Digital Simulation of Random Processes and its Applications. *Journal of Sound and Vibration*, **25**, 111–128, 1972.

119. Silva, W.: Soil Response to Earthquake Ground Motion. Technical report, Electric Power Research Institute, Palo Alto, USA, 1988.

120. Simmer, K.: *Grundbau 1: Bodenmechanik und erdstatische Berechnungen*. Teubner Verlag, Stuttgart, GER, 19th edition, 1994.

121. Sindowski, K.-H.: Das Quartär im Untergrund der Deutschen Bucht (Nordsee). *Eiszeitalter und Gegenwart: Jahrbuch der Deutschen Quartärvereinigung*, Vol. 21, Verlag Hohenlohesche Buchhandlung, Ferd. Rau, Öhringen, GER, 33–46, 1970.

122. Society for Industrial and Applied Mathematics: Linear Algebra PACKage. http://www.netlib.org/lapack/, Philadelphia, USA, 2005.

123. Song, C.; Wolf, J.P.: Consistent infinitesimal finite-element cell method: out-of-plane motion. *Journal of Engineering Mechanics ASCE*, **121**, 613–619, 1995.

124. Song, C.; Wolf, J.P.: consistent infinitesimal finite-element cell method for diffusion equation in unbounded medium. *Computer Methods in Applied Mechanics and Engineering*, **132**, 319–334, 1996.

125. Song, C.; Wolf, J.P.: Consistent infinitesimal finite-element cell method: three-dimensional vector wave equation. *International Journal for Numerical Methods in Engineering*, **39**, 2189–2208, 1996.

126. Song, C.; Wolf, J.P.: Consistent infinitesimal finite-element cell method for incompressible unbounded medium. *Communications in Numerical Methods in Engineering*, **13**, 21–32, 1997.

127. Stamos, A. A.; Estorff, O. v.; Antes, H.; Beskos, D.E.: Vibration isolation in road-tunnel traffic systems. *Int. Journal for Engineerng Analysis and Design*, **1**, 109–121, 1994.

128. Steinbach, O.: Vorlesung Hierarchische Matrizen. Berichte aus dem Institut für Angewandte Analysis und Numerische Simulation, 2004. http://preprints.ians.uni-stuttgart.de.

129. Streif, H.: *Geologische Karte von Niedersachsen 1: 25000, Erläuterungen zu Blatt Nr. 2314 Hooksiel.* Niedersächsisches Landesamt für Bodenforschung, Hannover, GER, 1985.

130. Taylor, G.J.: The spectrum of turbulence. *Proc. R. Soc.*, **Ser.A 164**, 476–490, 1938.

131. Thompson, L.L.; Pinsky, P.M.: *Acoustics*, Vol. 2, *Encyclopedia of Computational Mechanics*, Chapter 22, 695–717. Wiley and Sons, Chichester, UK, 2004.

132. Tsynkov, S. V.: Numerical solution of problems on unbounded domains. A review. *Applied Numerical Mathematics*, **27**, 465–532, 1998.

133. Turkel, E.; Yefet, A.: Absorbing PML boundary layers for wavelike equations. *Applied Numerical Mathematics*, **27**(4), 533–557, 1998.

134. United Nations: UN Atlas of the Oceans, Energy Resources. http://www.oceansatlas.com/unatlas/-ATLAS-/chapter8.html, 2005.

135. University of Washington: Soil Liquefaction. http://www.ce.washington.edu/ liquefaction/html/quakes/kobe/kobe.html, Seattle, USA, 2005.

136. U.S. Department of Energy: History of Wind Energy. http://www.eere.energy.gov/windandhydro/wind_history.html, USA, 2005.

137. U.S. Geological Survey: International Registry of Seismograph Stations. http://neic.usgs.gov/neis/station_book/, Washington D.C., USA, 2005.

138. van Wijk, A.J.M.; Beljaars, A.C.M.; Hotslag, A.A.M.; Turkenburg, W.C.: Evaluation of stability corrections in wind speed profiles over the North Sea. *J. of Wind Eng.and Ind. Aerodynamics*, **33**, 551–566, 1990.

139. Veers, P.S.: Three-Dimensional Wind Simulation. Technical report, Sandia National Laboratories Report, SAND88-0152, Albuquerque, USA, 1988.

140. Vibrationdata: Earthqukes. http://www.vibrationdata.com/earthquakes/kobe.htm, 2005.

141. WAFO group: Wave Analysis for Fatigue and Oceanography. http://www.maths.lth.se/matstat/wafo/, Lund University, S, 2005.

142. Wendel, K.: Hydrodynamische Massen, Trägheitsmomente. *Jahrbuch der schiffbautechnischen Gesellschaft*, **44**, 207–255, 1950.

143. Wiemann, J.; Lesny, K.; Richwien, W.: *Gründung von Offshore-Windenergieanlagen – Gründungskonzepte und geotechnische Grundlagen.* Verlag Glückauf GmbH, Universität Essen, GER, 2002.

144. Wilson, J.F.: *Dynamics od Offshore Structures, 2nd ed.* Wiley and Sons, Hoboken, USA, 2003.

145. Wind Service Holland, WSH: Offshore Wind Energy. http://home.wxs.nl/ windsh/offshore.html, Oosternijkerk, NL, 2005.

146. Winkelaar, D.: SWIFT - Program for three-dimensional wind simulation, part 1: model description and program verification. Technical report, ECN, 1992.

147. Wolf, J. P.; Song, C.: *Finite-Element Modelling of Unbounded Media.* Wiley and Sons, Chichester, UK, 1996.

148. Wolf, J.P.: *The Scaled Boundary Finite Element Method.* Wiley and Sons, Chichester, UK, 2003.

149. Wrobel, L.C.; Aliabadi, M.H.: *The Boundary Element Method.* Wiley and Sons, Chichester, UK, 2002.

150. Yang, Z.; He, L.; Bielak, J.; Zhang, Y.; Elgamal, A.; Conte, J. P.: Nonlinear Seismic Response of a Bridge Site Subject to Spatially Varying Ground Motion. In *16th ASCE Engineering Mechanics Conference.* (Turkiyyah, G. et al., Ed.), University of Washington, Seattle, USA, 2003.

151. Zhang, Y.; Yang, Z.; Bielak, J.; Conte, J. P.; Elgamal, A.: Treatment of Seismic Input and Boundary Conditions in Nonlinear Seismic Analysis of a Bridge Ground System. In *16th ASCE Engineering Mechanics Conference*. (Turkiyyah, G. et al., Ed.), University of Washington, Seattle, USA, 2003.

152. Zienkiewicz, O. C.; ; Chan, A. H. C.; Pastor, M.; Schrefler, B. A.; Shiomi, T.: *Computational Geomechanics with special Reference to Earthquake Engineering*. Wiley and Sons, Chichester, UK, 1999.

153. Zienkiewicz, O.C.: *The Finite Element Method*. McGraw-Hill, New York, USA, 1977.

Index

Printing: Krips bv, Meppel
Binding: Stürtz, Würzburg